Top100

品味 TOP100
TASTE

编著

让中国人住得更美好

U0253041

海峡出版发行集团 | 福建科学技术出版社
THE STRAITS PUBLISHING & DISTRIBUTING GROUP | FUJIAN SCIENCE & TECHNOLOGY PUBLISHING HOUSE

"和"与"美"：高品质生活

高品质生活是一种追求品位、舒适的体现，同时也是一种博雅的情怀，更是一种静水深流的境界。它强调品味、品鉴，注重外观和内涵的统一协调，不仅仅只是表象，也是自然内涵的流露，更是对生活要有一种成熟理性的思想认识。

虽然每个人的生活状态和认知都可能存在差异，但是"高品质生活"的"品质"理应是统一的，它是一种以"和""美"为核心特质的理想生活状态。

"和"是中国传统哲学的一个重要范畴，也是中华文明的追求目标。在对"高品质生活"的追逐，"和"就是对人与人、人与自然和合同一关系的追求。概括来讲，所谓的"高品质生活"，就是能够正确处理一切生活关系的一种状态。

那么，如何正确处理这一关系呢？这里，我们可以通过"高品质生活"的第二个特质——美。费孝通先生的"各美其美，美人之美，美美与共，天下大同"，虽然是对文化观的表述，但也完全可以作为生活中应遵循的原则。

人有千百样，生活自然也有千百种。各美其美，就是要承认差异，懂得欣赏他人和他人生活之美。美人之美，则是在各美其美的基础上，尊重、包容他人的个性与生活美。最后的"美美与共，天下大同"，就是以善意交换善意、以美好触碰、融合美好，以此孕育新的美好，最终实现"天下大同"的理想。只要做到以上几点，自然就实现了自我与自我、自我和他人、社会的和谐统一、共荣共生；当身心和、家庭和、社会和的时候，自然就有了好的生活体验，生活品质也自然得以提高。

"高品质生活"不仅是高品质、美好的人、人与人关系，也必然包含艺术审美方面的内容，特别是人与生活、自然环境关系。相对于"和"，"美"则突出表现在因物质生活富足，所带来的美好生活体验之上。这一体验，也是建立在诸如美食、美景、美物、美器等一切利于提升生活品质的物品之上。物质之美，同样是千人千面千好。

正如前文所言，"高品质生活"是"和"的、"美"的，是精神生活与物质生活共茂、身体与灵魂共荣的理想，更是我们人生理所当然的追求目标和向往。

目录 / CONTENTS

006 评委 JUDGES

008 陈设艺术空间 FURNISHINGS ART SPACE

096 别墅空间 VILLA SPACE

182 公寓住宅空间 RESIDENTIAL SPACE

270 创意设计空间 CREATIVE SPACE

JUDGES
评 委

终审评委
Special Support

吴宗敏
广州大学美术与设计学院党委书记，教授，硕士生导师

余静赣
联合国国际生态生命安全科学院院士

刘晓萍
华浔品味装饰集团创始人之一、全国总设计师

胡小梅
广东省陈设艺术协会执行会长

谢英凯
中国建筑学会室内设计分会理事会副理事长

刘明富
华浔品味装饰集团高级副总裁

叶玲菊
新浪家居华南大区主编

Antonio Lamonarca
Antonio Lamonarca 建筑事务所创始人

Maria Materia
意大利米兰 Lamonarca 建筑室内设计事务所合伙人

初审评委
Special Support

吴文华
华浔管理设计学会泉州分会

何志潮
华浔管理设计学会佛山分会

吴旭东
华浔管理设计学会江西分会

邱欣林
华浔管理设计学会重庆分会

卓谨仲
华浔管理设计学会东莞分会

叶 翔
华浔管理设计学会重庆分会

邹广明
华浔管理设计学会江苏分会

聂 雨
华浔管理设计学会广西分会

付调辉
华浔管理设计学会湖南分会

FURNISHINGS ART SPACE
陈设艺术空间

2020 / 品味 TOP 100

010	暗香盈袖 / FAINT FRAGRANCE PERVADES SLEEVES	056	恒利·江悦明珠148 / HENGLI·JIANGYUE MANSION 148
014	联发君玺 / ROYAL SEAL	058	瑞和公馆 / RUIHE HOMES
018	万达俊豪世家 / ELITE FAMILY	062	满堂院白色 / A WHITE HOUSE
022	嘉福·尚江尊品 / PRESTIGE RIVERSIDE	066	碧桂园府前一号 / HOUSE NO.1
024	中式禅意风 / ZEN HOUSE	070	东海泰禾 / DONGHAI TAIHE
028	美丽沙天寰湾 / BEAUTIFUL SANDY BAY	074	幸福 N 次方 / N-DIMENSIONAL HAPPINESS
032	云府悦城 / OVER THE CLOUD	076	居然设计中心办公室 / THE DESIGN CENTER (OF EASY HOME)
036	用设计成就美好家 / EXCELLENT DESIGN MAKES SWEET HOME	080	东方花苑 / ORIENTAL GARDEN
038	寻觅 / SEEKS	082	新大陆一号 / NEW WORLD NO.1
042	九龙仓 / WHARF HOLDINGS	086	海润·棠颂 / HAIRUN·TANGSONG
046	打造有温度的生活空间 / A COZY LIVING SPACE	090	温柔以待,时间的答案 / BE GENTLE, THE ANSWER IS TIME
050	暨阳湖壹号 / JIYANG ROYAL GARDEN	094	约克郡 / YORKSHIRE
052	所谓·遇见 / ENCOUNTER		

暗香盈袖

FAINT FRAGRANCE PERVADES SLEEVES

项目名称　皇庭丹郡
设　　计　华浔品味装饰 / 何俊杰
建筑面积　183 平方米
项目地址　福州
主要材料　奥特曼大理石、新中源瓷砖、进口墙布、进口硬包
工程施工　华浔品味装饰

在这个充满数据化的喧嚣年代，居家空间比以往任何时候都更能证明它是一个能够提供美好生活的载体。本套业主有着留学海外的背景，美式轻奢格调将其对居住的憧憬落到了实地，并以一种平衡的姿态让空间拥有了强大的亲和力，调高，却不离众。

在设计师的理解中，"奢华"作为一种表现载体，它应该是经久耐看的舒适淡雅，而不是喧哗纷繁的高调张扬。尤其对于年轻人的居所，这种对奢华的再定义更是不能规避的设计命题，于是在视觉通透的客厅中，将美式家具融入现代的居住模式中。

In this age of data, home space has, more than ever been regarded as a carrier for a better life. Having studied abroad, the owner chose the American accessible luxury style to build his ideal home. And this space has acquired a strong affinity in a balanced way.

In the understanding of the designer, luxury is taken as a carrier of expression, it should be durable and elegant, rather than high-profile, especially for young people's residences, the redefinition of luxury is a task that cannot be avoided in design. Therefore in the visually transparent living room, American style furniture is integrated into the modern life pattern.

笔触不多的金色细节装饰，以适度的奢华质感放大空间的感官张力。目之所及，或浓或淡的色调与材质的轻重混搭贯穿整个空间，使得风格与生活方式产生密切的联系和渗透，将一份关于品质生活的舒适铺洒在周遭。

餐厅与客厅呈开放式布局，清浅的色调与舒朗的格局让空间透出优雅的气质，不经意间放慢了生活的节奏。这种含蓄、考究，亦不乏温厚之感，或许是当下人们对于居住空间最直截了当的需求。在这样轻盈的环境中，橘黄色的餐椅则让这方区域有了些许"落地生根"的踏实，戏剧化的设计让居者在空灵、明朗中丰富了对浪漫主义风格的理解。

The small amount of golden ornaments, enhances the sensory tension of the space with its moderate luxury. As you can see, light and heavy shades and materials are combined throughout the space, infiltratingthe style into the lifestyle, and spreading around a comfort of quality life.

The dining room and living room are open-plan spaces, with light tones and a bright layout that gives the space an elegant temperament and inadvertently slows down the pace of life. This subtlety, exquisiteness and warmth, might be the most straightforward demand for living spacenowadays. In such a light environment, the orange dining chair adds a little down-to-earth feeling to the area, and the dramatic design enriches the owner's understanding of romanticism in the ethereal and clear air.

联发君玺

项目名称	联发君玺
设　　计	华浔品味装饰 / 赖斌、段小慧
建筑面积	140 平方米
项目地址	赣州
主要材料	木饰面、不锈钢、大理石、哑光地砖、硬包、墙布
工程施工	华浔品味装饰

ROYAL SEAL

客厅采用的是当下正流行的无主灯设计，迎合极简空间的断舍离，让空间更显高级感，同时也满足室内灯光需求。整体空间采用高级灰调加中色调，营造宁静的居家氛围。简约不等于简单，一桌一椅设计感都很强。白色大理石背景墙在家装是比较常见的，大气，简洁，优雅，符合大部分人的审美。

餐厅整体色调与客厅完美契合，空间得到延续。主卧空间也呈灰色调，空间达到了一致的统一。衣帽间玻璃推拉门，突显高级感，床头柜的设计是一个点睛之笔，既漂亮又省空间。

The fashionable deconcentrated lighting is elegantly luminating the living room, echoing with the minimalist space .The whole space is tuned with balanced gray tone and neutral colors, creates a serene atmosphere. Simple but not boring, the chairs, tables…everything in this space are well designed. White marble wall panal is more familiar to chinese audience, simple and refined , in line with most people's aesthetic taste.

The dining room and the sitting room assembled a tonal harmonic, continuance space. The master bedroom is also in grey tone, consistent with the whole space. The walk-in closet has luxury sliding doors made of glass, the ingeniously designed bedside cabinet is both aesthetic pleasing and space saving.

男孩房以浅蓝色为主，桌椅与窗帘色调相呼应，加上装饰画一抹橙色互补，画面感不沉闷，显得欢快，舒适。

女孩房第一眼的感觉温馨、梦幻、甜蜜，以粉色系为主，每个女孩从小心中都有一个甜美浪漫的公主梦。

The boy's room is toned in blue, desk and chairs and curtains echo each other in color, a dash of orange in the decorative painting adorns the whole place, giving a cheerful, relaxed atmosphere.
The girl's room is in pink, giving a warm/ sweet/ dreamy feeling, catering to every little girl's romantic princess dream.

万达俊豪世家
ELITE FAMILY

项目名称 万达俊豪世家
设　　计 华浔品味装饰
建筑面积 150平方米
项目地址 海南
主要材料 大理石、黑台钛、硬包、FU实木定制
工程施工 华浔品味装饰

本案设计师以灰白色、咖啡色、黑色为主基调，深海蓝和绿色为点缀色，通过设计元素、色彩、照明、原材料的简化，简单明快的线条，恰到好处的点缀，营造出简单又随心的生活空间，让回家的人，身体和心灵都能回归到最为舒服的状态，在真实而静美的空间里活出最真实的样子！

房子南北通透布局较为合理，不足的是内部使用面积偏小较为局促，空间感不强，功能空间划分略差，入户地方尤其需要做调整。设计师通过调整处理，功能空间更加宽敞，使用也合理，又能最大限度地保证使用度和空间感。

The decorator toned the house with grey, coffee color, and black, dotted with navy-blue and green. Reduced decorations, integrated lightings, simplified elements, clean lines, construct a simple and relaxing living space, make the most comfortable home, a soothing space for all the family members!

The house has a reasonable layout, southward and northward windows promoting a nice cross-ventilation. The original internal area was not spacious enough, space partition was not satisfactory. So our decorator made a few adjustments to create a more capacious space with higher efficiency.

尽可能不用装饰和取消多余繁冗的陈设摆放物品。在空间的装饰上，任何复杂的设计，没有实用价值的特殊部件及任何装饰都会增加空间的紊乱，但在细节部位要稍作修饰，让整个空间会感觉清新活跃色彩鲜亮。

All the redundant decorations and displays are canceled. When it comes to space decoration, any complex design, useless components and decorations could result clutter, but a few effortless details will light up the whole space and give a fresh and bright feeling.

嘉福·尚江尊品

PRESTIGE RIVERSIDE

项目名称	嘉福·尚江尊品
设　　计	华浔品味装饰 / 洪文龙
建筑面积	258 平方米
项目地址	赣州
主要材料	大理石，不锈钢，硬包，墙布
工程施工	华浔品味装饰

温柔，是这套作品给人的第一感觉，你来或不来，她就在那里，不悲不喜。如果温柔有颜色，那一定是莫兰迪色，不争不抢，不谙世俗，以舒缓自然的姿态直抵人心。

整体颜色干净纯粹，没有大面积的亮色，也没有夺目的色彩点缀，所有装饰都安安静静地在那里，仿佛一幅仕女图，却舒服得让你离不开眼。所有用色不张扬，却能相互制约，相互抵消，让视觉达到完美平衡……

最普通的颜色组合，却被赋予了高级的灵魂。安安静静做自己，莫不是如此。

Tenderness, is the first impression this work gives.

You come or not, she is there, neither sad nor happy. If there is a palette for tenderness, it must be the colors of Morandi, reach your heart directly with a soothing calming attitude, no need to rush and hush.

The palette is clean and pure, without extensive bright color, without dazzing adornment, everything is just there, quietly, easily, just like a picture of a fine lady. All the colors are subtilely harmonic, perfectly balanced...

The most common color combination, with the most advanced soul. Just be yourself, quietly.

中式禅意风
ZEN HOUSE

项目名称 金域蓝湾
设　　计 华浔品味装饰
建筑面积 100 平方米
项目地址 广州
主要材料 地砖、麻织地毯、实木地板、乳胶漆、新中式家具
工程施工 华浔品味装饰

一盏茶，一卷书，独坐晨光里，
任窗外高楼林立，车马喧嚣，
我自漫步于心中的禅房，
轻轻缓缓，不争，不言，
只有把一颗心完全地交付出去。
主人爱好茶艺，设计师结合业主的喜好与生活习惯在阳台空间规划了韵味十足的中式茶室区域。几缕柔和的晨光或日落之前的橙黄色的霞光，一张茶几，一把椅子，再一盏茶，打在地上的斜斜的影子，便构成一副绝美而宁静的画面。

Sitting alone in the morning light, With a cup of tea and a book,
Care not how high the buildings rising and how busy the cars moving outside the window,
I walk in the Zen house of my heart,
Gently and slowly, without hustle, without words,
Just give your heart out.
The decorator designed a charming chinese style tea house on the balcony for the tea loving host.
A tea table, a chair, and a cup of tea, in the soft morning light or the orange glow before sunset,
with a slanting shadow hitting the ground, make an extremely beautiful and serene scene.

美丽沙天寰湾
BEAUTIFUL SANDY BAY

项目名称　美丽沙天寰湾
设　　计　华浔品味装饰 / 肖晨曦
建筑面积　300 平方米
项目地址　海口
主要材料　原木家具、乳胶漆、原木墙板、简一瓷砖
工程施工　华浔品味装饰

对于在繁忙都市中的我们而言，原木色家居所营造的闲适写意、悠然自得的生活境界，如一股清流，缓缓地在我们心间流淌，让人想闭上眼睛，享受那份难得的宁静。严谨安静的色彩，让空间展现出优雅干净之态，颇有东方禅思禅韵。

For people living in busy modern cities like us, burlywood color suggests a leisure and carefree lifestyle. The burlywood colored dwelling feels like a clear current, flows slowly in our heart, makes you just want to close your eyes, enjoy the peaceful time.
Sober calming colors, give the space a clean and elegant feeling, with a hint of oriental charm and Zen spirit.

OVER THE CLOUD

云府悦城

项目名称 云府悦城
设　　计 华浔品味装饰 / 曾一晟
建筑面积 117 平方米
项目地址 赣州
主要材料 大理石、护墙板、墙布、硬包、不锈钢
工程施工 华浔品味装饰

美，是愉悦开心的源头。

本案体现出来的是主人对幸福的憧憬和美好的想象，没有复杂的设计，却处处彰显着品位，让人眼前一亮。整体入户柜的设计让空间整洁不凌乱，收纳空间充足，极大地利用了空间；中部留空，美观又实用，时常放些使用率较高的物件，方便快捷；出门之前对着镜子整理仪容，自信心满满，心情愉悦一整天。

Beauty is the source of happiness.

This work is the projection of the owner's prospects and imaginations. Without complex design, it is refreshingly tasteful. Integrated cabinet by the entrance is both pretty and practical, provides ample storage space, keeps the whole space organized and tidy. All the high usage stuff can be stored here, very convenient. There is a mirror in the doorway, so the inhabitants can dress smartly and walk out confidently and happily.

米白基调的空间在灯光的烘托下，满满的精致、温馨、浪漫和优雅。家居大气不过于华丽，完美地营造出轻奢的小资格调，舒适、气派、实用和多功能。空间柜体颜色款式元素相近，基本一致，空间延续性强。

酒柜与电视背景墙并齐，收纳空间满满，背景墙壁灯与之呼应，氛围感和精致感呼之欲出。电视柜和茶几采用的是大理石台面，大气简洁，易清理。金属的点缀让空间有了轻奢的质感。

With perfect lighting, the beige toned space looks refined, warm, romantic and elegant, makes a comfortable, luxurious, practical and multi-functional petty bourgeoisie home.The color and style of the furnitures in the space are similar and harmonious, spatial continuity is strengthened. Wine cabinet and TV background wall are of the same height, the wine cabinet provides sufficient storage space, and echoes with the background wall in style, creating an exquisite feeling. The TV ark and tea table are made of marble, with clear outlines and will be easy to clean. The metal ornaments give the space more luxury touches.

用设计成就
美好家

EXCELLENT DESIGN MAKES SWEET HOME

项目名称　南海幸福城
设　　计　华浔品味装饰 / 杨小明
建筑面积　160 平方米
项目地址　海口
主要材料　大理石、不锈钢、墙布、乳胶漆
工程施工　华浔品味装饰

本案在色彩选择上，选用了带有高级感的中性色，诸如象牙白、奶咖、黑色、黄色及高级灰，来演绎一种"低调的奢华"。借助亮色系的挂画、布艺、抱枕、家具等，为空间增添一抹亮丽的同时提升了鲜活感，让整个室内空间质感更为饱满，呈现温馨大气的格调之余，又能让人感受到浓浓的时尚高雅范儿。

客厅以白色、黑色和高级灰为主色调，黄金色、宝蓝色为辅助色，配上精挑细选出来的艺术品，每个细节都透露着精致优雅。顶面无主灯设计，全屋使用简约的几何线打造出空间的层次感。地面是大理石花纹的瓷砖，简约时尚，给人干净利落的感觉。沙发背景墙大理石自然的纹路低调又奢华。电视背景墙的护墙板，造型简洁大方，却又提升了空间的时尚感。

Ivory white, coffee milk, black, yellow and grey ... such high class neutral colors are used in this design, make the palce looks modest but luxurious. Bright colored decorative picture, soft adornment, pillows, furnitures... give dashes of freshness, so the whole interior space looks warm and stylish.

The living room is dominated with white, black and high-class gray, interspersed with gold and royal blue, and decorated with carefully selected artworks, everything is exquisite. With deconcentrated lighting, simple geometric lines sketch up the contours of the space. The floor is paved with marbled tiles, both simple and stylish, looks pretty neat. The marble wall behind the sofa looks casually luxurious with beautiful natural marble vains. The clapboard of the TV backdrop wall looks clean and chic.

寻觅
SEEKS

项目名称 保利城
设　　计 华浔品味装饰 / 俞欣
建筑面积 90 平方米
项目地址 泉州
主要材料 大理石、硬包、不锈钢
工程施工 华浔品味装饰

本案属于轻奢风格，没有过度浮夸的装饰，也没有多余的装饰，它的魅力在于恰到好处的"奢华"，在"简"与"奢"之间寻找平衡。洁白的大理石茶几充满了高级感，与同样材质的餐桌互相映衬，在金属吊灯的配合下丰富了室内元素。

This is a Casual Luxury space, without excessive or redundant decoration, its charm lies in the fine place between "simplicity" and "luxury". White marble tea table is totally classy, echoing with the dining table made of the same material, shined by the metal chandelier enriching the interior elements.

WHARF HOLDINGS

九龙仓

项目名称 九龙仓
设　　计 华浔品味装饰 / 潘骏
建筑面积 120 平方米
项目地址 杭州
主要材料 高端灰、护墙板、玻璃、乳胶漆
工程施工 华浔品味装饰

舍弃浮华，造型简单，装饰有度为本案的主题。

优雅沉静的灰，搭配原木色元素将整体氛围烘托出来，在用材用色方面主张宁缺毋滥的设计理念，空间通过摆件、画品、灯光、以及各个细节处理，转换成空间中强烈的视觉对比效果。微妙的视觉感受让一切变得很和谐，宁静自然。

Abandoning ostentation, simple modeling and moderate decoration are the themes of this case.

Elegant and quiet gray, with log color elements to set off the overall atmosphere, in the use of wood color, advocate the design concept of "lack of no abuse". The space is transformed into a strong visual contrast effect in the space through the processing of ornaments, paintings, lighting, and various details. The subtle visual feeling makes everything harmonious, peaceful and natural.

打造有温度的生活空间

A COZY LIVING SPACE

项目名称　华侨城
设　　计　华浔品味装饰 / 刘云
建筑面积　80 平方米
项目地址　重庆
主要材料　原木、木地板、乳胶漆
工程施工　华浔品味装饰

屋主是三口之家，向往简单精致的生活环境，家里小孩较小，对衣、食、住、行的需求也要全部考虑。设计师用简单的形式，"以少为多"，通过对色调、线条、材料的灵活运用，营造一种云淡风轻的温暖气质。

北欧设计，经久不衰。在有限的空间里，设计师巧妙实现实木家具 + 暖色调的搭配，再以一张北欧系灰色皮质沙发呼应，视觉上轻快简约，带着一种自由放松的气息，用心诠释空间的故事。

The owner's family has three members, the kid is still young, the needs of a small kid's clothing, food, living and transportation should be taken into consideration. Designers used a simple form, created a cozy place to live, with smart use of colors, lines, materials. "less is more".

Nordic design is enduring. In the limited space, the decorator skillfully designed a relaxing space with solid wood furniture + warm colors. The Nordic style gray leather sofa fitted in effortless with simple design and light colors in this space.

木质家具的温润，让家在阳光里更温暖。在角落的绿植为空间增添了生命力，非常自然地融入家居设计中。无论是家具，还是灯饰，外型线条都很轻盈，在极简的线条外观下，打造出一种空间宽阔、内外通透的视觉感受。

北欧不一定就是"冷淡"、不解风情的，它也有柔情浪漫的一面。墨绿色背景墙给空间增加一丝活力，和煦的阳光洒在木质地板上，给地面铺上一层温暖，配搭北欧简单造型的家具，享受清爽干净的睡眠环境。

The wooden furnitures make the home feels warmer, especially in the sun. Green plants in the corner, naturally blend into the whole space, make the space feels more lively . The furnitures and lights have clean lines, and the minimalist lines structured this open and airy space.

Nordic design is not necessarily "cold" or "apathetic", it could be tender, even romantic. The dark green background wall adds a bit of vitality to the space. The sun shines on the wooden floor, spreading a layer of warmth to the ground. Equipped with simple nordic style furnitures, this is a fair and tidy sleeping environment too.

暨阳湖壹号

JIYANG ROYAL GARDEN

项目名称 暨阳湖壹号
设　　计 华浔品味装饰 / 彭楠
建筑面积 360 平方米
项目地址 张家港
主要材料 木地板、墙布、大理石
工程施工 华浔品味装饰

本案设计风格以轻奢为主，设计师对精致简约的欣赏和热爱，注定要将空间的统一性做到最好，而且要把细节和写意一一呈现。

低调却富有质感的金属元素，勾勒出一方典雅的轻奢生活空间。沙发的配色极具摩登复古感，实用与颜值并存。大理石的地面搭配绒毛地毯，营造出空间高雅、奢华以及大气的氛围。现代风和古典风完美的结合，凸显出房主的时尚品位。精致复古的物件，加入花卉设计装饰，使得整个空间充满生气、活力，单调的空间而不乏清新的气息，享受悠然与轻松的现代家居生活和高品质的生活方式。

This is a typical casual luxury design. The decorator is inclined to make refined minimalist design, impeccable harmonic place with fine details.

The modest but classy metal elements outlined this casual luxury living space. The color of the sofa has a mixed modern-retro style, both practical and dashing. The marble flooring and fluffy carpets create an atmosphere of elegance, luxury and grandeur. The perfect combination of modern and classic demonstrated the extraordinary taste of the owners. Exquisite retro style objects, together with floral decorations, make the otherwise monotonous space full of vitality. The only thing left is to enjoy the high quality leisure life in the modern time.

所谓·遇见

ENCOUNTER

项目名称　所谓·遇见
设　　计　华浔品味装饰 / 林凯平
建筑面积　139 平方米
项目地址　福州
主要材料　道格拉斯瓷砖、唯美LD瓷砖、雅士白大理石、芬琳漆、绎尚巴伦窗帘
工程施工　华浔品味装饰

所谓遇见，不可求，不可赊。

和女主人的遇见是一个阳光的午后，没有早、没有晚的不期而遇，短发、干净素色的着装、淡淡的微笑。也许就是这一抹微笑，她未来的家就在我脑海一跃而出。

简约、自然、没有多余的装饰，通过大理石优美的纹理和细腻的木饰面，让空间变得更加精致有层次感。实木的融入，让大理石多了几分朴素温暖的韵味。

Encounter, is something that can not be forced, nor expected.

I met the owner on a sunny afternoon. She is short-haired, wearing clean plain color clothesand a smile. Perhaps it was just from this smile, her future home designcame to mind.

The project is simple, natural and without excessive decorations. Thegraceful texture of marble and delicate wood finish enhance the refined atmosphere and layering of space. The solid wood adds a little warm and simple impression to the marble.

两种材质的完美融合,将功能与美学转化为精致的细节,赋予其刚柔并济的空间氛围。它温润而自然,低调而内敛,看似清冷的姿态,却流露出时尚奢华的质感。有木则雅,有石则贵。

The perfect blending of these two materials turns functions and aesthetics into delicate details, creating a space that is both rigid and soft. Being gentle, natural and understated, the seemingly cold space is revealing an aroma of fashion and luxury. The space looks elegant with the wood elements, and grand with the marbles.

恒利·江悦明珠148
HENGLI · JIANGYUE MANSION 148

项目名称 恒利·江悦明珠148
设　　计 华浔品味装饰 / 李波
建筑面积 148平方米
项目地址 泸州
主要材料 大理石、护墙板、诺贝尔瓷砖、实木
工程施工 华浔品味装饰

"敬、寂、和、清"源于茶道精神，也是儒家思想文化的体现。我们通过提取演变，以"敬、寂、和、亲"作为设计主题，使之更好地融入到本案的设计中去，从而营造一个让人心情愉悦的空间。敬，是万物之本，敬乃尊重他人，对己谨慎。尊敬：对人的尊敬；对传统文化的尊敬；对设计的尊敬。首先采用柔软的材质，对老人和小孩的保护，体现更人性化的设计；其次融入传统元素，是对传统文化的传承和敬重；从而延伸对设计的尊重——用心设计，更好地营造高品质空间。

Respect, serenity, harmony and purity originated from the spirit of tea ceremony and is also the embodiment of Confucianism. By extracting and evolving, we use "respect, silence, harmony, and family" as the theme to integrate them into this project, thus creating a delightful space.

Respect is the foundation of everything; it is to respect people, traditional culture and design. The selection of soft materials to protect the elderly and children. And traditional elements are integrated to show the inheritance and respect for traditional culture, building a high-quality space.

RUIHE HOMES

瑞和公馆

项目名称 瑞和公馆
设　　计 华浔品味装饰
建筑面积 156 平方米
项目地址 烟台
主要材料 仿大理石砖、护墙板、不锈钢
工程施工 华浔品味装饰

华灯初上，忙碌了一天的男主人回到家推开门映入眼帘的是一面整洁大气的大理石主题墙，在门厅把一天的疲惫"归置"好。客厅柔软的沙发在皮质背景的衬托下格外温暖，夹杂的墙面顶面的金属修饰让眼球也有些许的兴奋。

孩子们在书房已经把功课温习完毕，妻子精心准备的晚餐已经就绪，餐桌上的美食在橙红色的背景的映衬下挑动着味蕾。洁白的大理石餐桌面，金色的桌腿，蓝色的柔软适中的餐椅，精心挑选的装饰品，无不让一天的疲惫消失殆尽。

This project is positioned as a high quality living space for an elite couple and their two children.
At the evening arrives, the busy husband returns home and upon opening the door,he is facing a clean and tidy marble wall. After leaving his exhaustion in in the entranceway, the soft couch in the living room appears exceptionally warm against the leather finish in the background. The slight metal finish at the top of the wall also catches his eyes.

The children have finished their homework in the study, and the wife has prepared a nice dinner which looks appealing against the orange-red background. And the white marble table, golden table legs, blue softdining chairs, and carefully selected decorations, have completely cleaned away the exhaustion.

餐后多功能房是夫妻二人与孩子们沟通交流的一方小天地。爸爸妈妈席榻而坐，孩子们窝在爸爸的臂弯里开心地分享当天的课业，妈妈笑着给他们讲睡前的故事。又到了互道晚安的时候，哥哥依依不舍地回到属于自己的房间，软软的一方小床让他有足够的安全感，床头也是他最喜欢的画作。稚拙的童真，绿色是属于他的活力和自信，米色又是他需要的属于妈妈的温暖。

夫妻二人安顿好了孩子，男主人坐在书房，随手从书架上拿起一本书，这是他最喜欢的节奏，金色的书架黑色的层板是他的自信也是他对事业的追求，书桌对面墙上的画作是他多年的珍藏。

夜深人静，走过缓缓的并不局促的过廊来到的主卧室，木色、浅咖啡色的皮质硬包、透明的衣帽柜、米色的床品构成了这个独属空间，温暖、静谧、充满爱意，顶面点缀的玫瑰金，床头的叶片装装饰又给予了她独属的轻奢主题。

After dinner, the multi-function room is a small world where the couple communicates with their children. The parents are sitting on the couch and the children are happily sharingthe lessons of the dayin their father's arms. The motheris smiling and tellingthem a bedtime story. Then when it comes to bedtime, the brother reluctantly returns to his own bedroom, the soft cot gave him a sense of security. On the bedside is his favorite painting full of childish innocence.Green is the color for his vitality and confidence and beige stands for the warmth that he needs from his mother.

After putting the children to bed, the husband sat in the study and picked up a book from the shelf. This is his favorite rhythm. The golden shelf and black layer board is the confidence and pursuit of career. The painting on the wall facing the desk is his collection for many years.

In the dead of night, walk through the corridor to arrive at the master bedroom, the wood color, light coffee wall finish, transparent closet and beige bed sheets constitute this unique space, which is warm, quiet and full of love. while the rose gold embellishment on the ceiling and the leaf ornament at the bedside gives a distinctive luxury theme.

满堂院白色
A WHITE HOUSE

项目名称 满堂院白色
设　　计 华浔品味装饰
建筑面积 120平方米
项目地址 无锡
主要材料 乳胶漆、锈钢、玻璃、护墙板
工程施工 华浔品味装饰

屋主是一家三口——一对高颜值小夫妻，和一个可爱的女儿。女主人想要家里有华丽的线条，家具也要有精致的造型；男主人是典型的"直男"，偏好现代简约的风格，家居软装上要凸显质感。结合二者的意见，本套案例着重在色彩的运用及家具的选用上下功夫，塑造出一个气质的空间。

The owner of the house is a family of three - a good looking couple and a lovely daughter. The wife wants luxuriant lines and exquisite furnitures; The husband is a typical "straight guy", who prefers to keep modern design and simple lines, use soft adornments to decorate the house. Combined the opinion of both owners, this case is more about colorific apply and fitting in exquisite furnitures, to make the space special.

碧桂园府前一号
HOUSE NO.1

项目名称 碧桂园府前一号
设　　计 华浔品味装饰 / 何建生
建筑面积 268 平方米
项目地址 衢州
主要材料 定制家具、不锈钢、乳胶漆、瓷砖
工程施工 华浔品味装饰

本案是现代港式风格。客厅暖灰色的沙发虽然属于冷色调，但有几分温暖的柔和感，搭配渐变色系地毯呈现的是含蓄内敛却又有包容力的美感。客厅落地窗的设计在开阔视野有足够采光的同时，增添了室内与外界的联系。室内设计潮流多以现代风格为主，大多色彩冷静、线条简单。现代港式风格在处理空间方面一般强调室内空间宽敞、内外通透，在空间平面设计中追求不受承重墙限制的自由。墙面、地面、顶棚以及家具陈设乃至灯具器皿等均以简洁的造型、纯洁的质地、精细的工艺为特征，强调形式应更多地服务于功能。

This is a typical modern Hong Kong style design. Although the warm-gray of the sofa in the living room supposed to be a cool color, it somehow gives a warm and soft feeling. The gradual colored carpet feels reserved, introverted but tolerate. The French windows in the living room enlarge the visual field, let in more natural daylight, and enhance the connection between the interior and the outside space. The interior design is mainly modern, featuring with calm colors and simple lines. Modern Hong Kong style in space design generally stresses on airy and spacious interior space, interaction of indoor and outdoor, and the architecture structure and layout should make sure the load-bearing walls will not be obstructions to realize the open and spacious design. Simple lines, pure materials, and fine craft are preferred on walls, flooring, ceilings and furnitures, even lamps. Form serves function.

东海泰禾

DONGHAI TAIHE

项目名称 东海泰禾
设　　计 华浔品味装饰 / 俞欣
建筑面积 1200 平方米
项目地址 泉州
主要材料 无缝瓷砖、护墙板、铁艺
工程施工 华浔品味装饰

本案设计根据客户需求，将工业风的设计理念完美融入，运用高质感材料，打造出一个将别具复古气息和粗犷风格相结合的传媒公司。随意休闲、轻松愉悦是本案的特点。空间上，整体布局上错落有致，办公区域与休闲区域合理分隔，大胆利用美岩板进行柱子装饰与雕刻，用数字的方式进行办公区域、洽谈区、休闲区的划分，不仅具有美感，而且能引导用户。地板上，利用创意地面，浅灰色格调与工业风完美搭配，保持了地面的完整性，使得整个办公区充满和谐氛围。

This case is designed to meet the needs of customers, integrate the design concept of industrial style, and use highly textured materials to create a media company that combines vintage and rugged. Leisure and relaxation are the hallmark of this case. In terms of space, the overall layout is well-spaced, office area and leisure area are reasonably separated, and the rock plate is used to carry out the decoration and engraving of the pillars. The division of the office area, the negotiation area and the leisure area by digital means is not only aesthetic, but also to guide users. On the floor, the use of creative floor PANDOMO, the light gray style and the industrial style perfectly match, maintaining the integrity of the ground, is making the entire office area full of harmonious atmosphere.

顶面水泥保持原有色彩，裸露的水电管道，体现了工业风，粗犷的特色，让人产生强烈的视觉冲击感，在灯光的照射下偶尔闪现出一道金属光，神秘感十足。墙面多使用白色艺术涂料，让人一进门感觉到干净利落，是对粗犷的工业风的补充。

The top cement maintains the original color, and the exposed water and electricity pipelines reflect the industrial style is not hidden, the rough features, creating a strong visual impact, and occasionally flashing a metallic light under the illumination of the lights, full of mystery. White art paint is used on the wall to make people feel clean and clean when they enter the door. It is to the style of industrial design complementary.

幸福 N 次方
N-DIMENSIONAL HAPPINESS

项目名称　华诚悦庭 10 号
设　　计　华浔品味装饰 / 兰明珠
建筑面积　195 平方米
项目地址　张家港
主要材料　定制家具、不锈钢、乳胶漆、瓷砖
工程施工　华浔品味装饰

本案以"幸福 N 次方"为题，将对幸福的多重感受与解读融入个性化的生活场景中，为客户呈现出释放幸福能量的生活美学空间，设计依托原有的硬装基础，融入简约柔和的色彩，打造休闲而个性的生活理想样态。

With "N-dimensional happiness" as the theme, the multiple feelings and interpretation of happiness are integrated into personal life scenes, this is a living space to recharge energy and happiness. Based on the original architecture structure, soft colors are blend in, created an ideal life of leisure.

居然设计中心办公室
THE DESIGN CENTER (OF EASY HOME)

项目名称	居然设计中心办公室
设　　计	华浔品味装饰 / 周应钦
建筑面积	500 平方米
项目地址	广州
主要材料	硅藻泥、不锈钢、玻璃、定制家具
工程施工	华浔品味装饰

全案设计中心力求突破、探索创新，对传统家装设计公司的办公理念提出"改革"的思路，率先将体验型的概念融入办公空间，力图为高端客户提供更轻松、时尚的现代洽谈、体验的空间氛围。设计师极具国际化的设计思维，为本案注入了当代最强的设计语言，对空间之间微妙结构关系拿捏到位，让空间功能得以出色演绎，让我们看到，这不仅是一个办公空间，更是传递设计人文、美学的载体。

The design center strives to break through, explore and innovate, put forward the idea of "reform" to the traditional ideas of interior design company, took the lead in integrating the concept of experience into the office space design, to provide more relaxed and fashional experience of business negotiation for the high-end modern customers.

With international design intellection & means, the designer used the strongest contemporary design language in this case. With good control of the subtle spatial relationship, the best function of every space is achieved. As we can see, this is not only an office space, but also a vehicle of humanity and aesthetics.

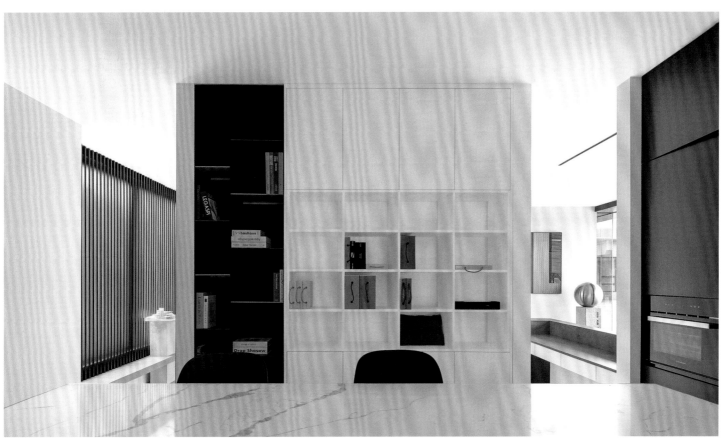

在对于这个220平方米的办公室的设计定位上,设计师在解决功能分布合理的基础上,着重打造如何在空间里展现企业文化,融入艺术美感,塑造全新的艺术生命及形态。空间设计采取开放式复合型,以期达到多场景切换、打造精致、轻松有趣的办公氛围,开放式的办公室也挥别了传统的格子间,营造出平等、合作、创新的工作氛围。空间采用"透"的设计手法,让空间彼此交互共存,产生出微妙的视觉趣味性,玻璃、铝条等现代材质带出了空间的透视和线条的次序感,光线的巧妙引入为素雅的空间注入了生命力。

In this 220 square meters office, on the basis of reasonable function division, the designer also stressed on demonstrating the corporate culture, integrating artistic sense of beauty into the new form and life. This space is an open compound of different components, ingenious and fun, versatile for different work scenes. Divergent from traditional cubicles, the open office created an equal, cooperative, innovative working atmosphere. Transparency is an important feature in this place, space can interact with each other, producing a subtle and delightful visual effect. Modern materials such as glass and aluminum strips gave depth and perspectives to the space, let the light flood in and bring out vitality to this simple and elegant place.

东方花苑

项目名称	东方花苑 59 号 901
设　　计	华浔品味装饰 / 汤海涛、程雅男
建筑面积	90 平方米
项目地址	张家港
主要材料	仿大理石砖、护墙板、乳胶漆
工程施工	华浔品味装饰

ORIENTAL GARDEN

本案在简约的基础上，加了一些轻奢的元素，黑白灰的色系搭配自然随性中带有一丝疏离感，传递"与世无争，我心安定"的家居理念。黑色的餐边柜时尚大气，搭配灰白花纹的大理石时尚不沉闷，创意吊灯增加空间设计感，整个餐厅既实用又美观。

This is a simple design, with a hint of luxury. With a sense of alienation, the house is decorated with black, white and gray, conveys the owner's attitude, "stand aloof from worldly affairs, my heart is settled".
The black sideboard with gray-white marble stripes looks elegant and chic, the innovative ceiling lamp completes this space with style, the whole dining room is both functional and pleasing to the eye.

新大陆一号
NEW WORLD NO.1

项目名称 新大陆一号
设　　计 华浔品味装饰 / 林凯平
建筑面积 189 平方米
项目地址 福州
主要材料 圣罗兰大理石，黑色墙布，米色窗帘，微晶石大理石，硬包皮革，黑钛，黄铜，科技木
工程施工 华浔品味装饰

家的丰富，生活的尺度，都离不开对空间结构与比例的考量，我们尽可能拆除必要之外的余赘之物，展现一处"进退"有度的场所。本案通过饰品、挂画、家具等组合搭配营造品位空间。每一件配饰、每一件生活用品、每一件器皿形成强烈的感官触觉符号转换到空间中，每个元素都蕴藏着卓然的审美品位，对空间产生一种近在咫尺的触觉价值体验。

We removed all the unnecessary redundance in this space so you have enough room for "advance and retreat". Considerate spatial structure and reasonable scale sustained the affluence of home and life.

Collocation of decorative pictures, ornaments and furnitures, creat a classy space. everything in this space contribute a remarkable aesthetic element and produce a valueable tangible experience to the space.

海润·棠颂

HAIRUN · TANGSONG

项目名称　海润·棠颂售楼部
设　　计　华浔品味装饰 / 罗枭、梁家豪、张大东
建筑面积　1500平方米
项目地址　晋江
主要材料　乳胶漆、仿大理石砖、硬包
工程施工　华浔品味装饰

本案从外观到内饰都采用了新中式风格。木材和石材是最常见的装饰，石材中又以大理石居多，灰白相间往往成为一幢售楼部的主色调，冷凛的色调让整体空间看起来意境深远。本案从外形整体看是偏江南风的设计，摒弃了现代风格的简约与单调，以及中国古典的繁琐，注重精美的设计细节，而且以独特的方式展现了"动"与"静"的环境。

This case is the sales centre of the real-estate developer Jinjiang Hairun Tangsong, both exterior and interior design and construction are undertaken by Huaxun decoration company. This is a Neo-Chinese style product from the inside out. Built with the most mainstream materials: Wood and stone, mostly marble. The gray and white color of the marble dominated the whole building, looking calm and poetic. The overall appearance of the design adopted the architectural form of traditional building from south of the Yangtze River. Without excessive architecture components of Chinese buildings, still different from the simple-monotonous modern style, the product balanced "dynamic" and "static" in a unique way with fine details.

楼盘模型区的水晶灯起着很好的渲染作用；背后的造型背景墙，灯光从墙上透出，更显高级。右侧是小区的地理位置图，为让客户都清楚楼盘的地理优势；右侧则是待客区，数张桌椅整齐排列，高达嵌入式储物柜用钛金条镶边再搭配灯条，突显奢华大气；帘子选用富有意境的水墨画。

The building model area is shined by the crystal chandelier and the background wall with embedded light bars. On the right side of model area is a map indicating the superior geographical location of estate; On the right is the reception area, a number of tables and chairs are arranged in order, cabinet with titanium edges and light bars looks luxurious; The curtain is featured with artistic Chinese ink and wash.

温柔以待，时间的答案

BE GENTLE, THE ANSWER IS TIME

项目名称　星城世家
设　　计　华浔品味装饰 / 林凯平
建筑面积　95 平方米
项目地址　福州
主要材料　瓷砖，德赛斯岩板，高光烤漆板，艺术涂料
工程施工　华浔品味装饰

开放式的客厅与餐厅相邻，吧台作为餐桌，如此非常规的布局方式，让相同的空间拥有不同的功能，跟家人的交流沟通毫无障碍。

橙色的单人沙发，白纱落地，沐浴阳光，营造出闲适温馨的阅读角落，精致的装饰体现出生活细节中的仪式感。

开放式的厨房，橱柜采用 U 形布局，合理的动线设计，下班回家看到爱人在厨房忙碌身影，让人想起一句话："最幸福的时光，从清晨的一碗清粥，到傍晚的一片云，在晨曦朝霞中挽手，在落日余辉中散步，相聚于三餐四季。"

The open living room is adjacent to the dining room, and the bar counter serves as the dining table. Such an unconventional layout enables the same space to have different functions, and unhindered communication among family members.

Orange single sofa, white gauze draped to the ground, the comfortable reading corner bathed sunshine… these are little ceremonies to celebrate life.

Open kitchen, cupboard with U-shaped layout, reasonable dynamic outlines, enable you to see your loved ones right there in the kitchen when you come home from work. As a saying goes : "the happiest time, comes from a bowl of fresh congee in the early morning, to a pretty cloud in the evening, from the hand-holding in the sunrise, to the walk in the afterglow of the setting sun, get together at three meals in four seasons."

主卧高级质感与利落的线条运用搭配,原本开放的主卧阳台,经过结构改造后摇身一变成为全景落地窗,开阔的视野让绝美的窗外景色一览无遗。

The master bedroom with neat outlines was decorated with high quality materials, the original open balcony is transformed into a panoramic French window, giving expansive vision of the stunning scenery outside.

约克郡
YORKSHIRE

项目名称	约克郡
设　　计	华浔品味装饰 / 唐陶
建筑面积	120 平方米
项目地址	重庆
主要材料	乳胶漆、瓷砖、艺术涂料
工程施工	华浔品味装饰

什么样的家，让人想要久待？公共区域明亮通风，假日也想待（呆）；通透大厨房＋餐厅，凝聚全家人的胃和心，在家就能聊天、喝下午茶；随手可拿的书籍！享受每一寸绿色自在的时光"情感交流"为出发点，创造巩固家人情感的场景，让户外自然景观与室内空间进行互动！

房间面积要小，要有景色，上天下地，惜山面海，每天回家看看不一样的云，听雨赏花，闻香，想想昨晚的梦，和自己聊一会天，日子容易丰盛起来。

In what kind of home, you would like to stay longer? The public areas should be bright and airy, so we will want to stay there on holidays. The kitchen & dining room should be ample and bright, so the whole family can have a good meal and a good chat, even have an afternoon tea at home; Books at hand! There should be "Emotional communication" amenities, both outdoor and indoor space for family interactions!

The room area should be small, there should be sceneries, of sky and grass, mountains and sea. Go home every day to see the different clouds, listen to the rain and enjoy the flowers, talk to yourself for a while, and think about your dreams. Days will be easy and sumptuous like this.

VILLA SPACE

别墅空间

2020 / 品味 TOP 100

098	蓝湖郡 THE BLUE LAKE		140	著美·印记 BEAUTY·IMPRINT
102	柔软的天 SOFT SKY		144	探索自由奥秘 THE FREE SPIRIT
106	东氿壹号 DONGJIU GARDEN		146	留白 THE BLANK SPACE
110	静美雅墅 SERENITY GRACE VILLA		150	星洲湾别墅 XINGZHOUWAN VILLA
112	心地自清幽 A PEACEFUL HEART AT A PEACEFUL RESIDENCE		152	裕顺雅苑 YUSHUN GARDEN
116	重塑理想生活 THE IDEAL LIFE		156	云星公园大观 YUNXING ROAMING WONDERLAND
118	欧风东韵 EUROPEAN GLAMOUR WITH ORIENTAL CHARM		160	南航——海岸华墅 COSTAL VILLA
122	英雄联盟 LEAGUE OF LEGENDS		162	九里峰山 JIULI FENGSHAN
124	居逸·一江光华 A RIVER OF BRILLIANCE		166	远洲墅 YUANZHOU VILLA
128	新强苑 XINQIANG GARDEN		170	蓝色多瑙河 THE BLUE DANUBE
130	缘园人家 MODERN ORIENTAL RESIDENCE		174	名门壹品 EMINENCE
134	阑——霖峰壹号 LINFENG GARDEN		178	让岁月留香来日可期 FRAGRANT DAYS
136	香缇半岛 NOBLE PENINSULA GARDEN			

蓝湖郡
THE BLUE LAKE

项目名称 蓝湖郡别墅
设　　计 华浔品味装饰 / 刘云
建筑面积 680 平方米
项目地址 重庆
主要材料 丝绒、皮饰、瓷砖、木饰面等
工程施工 华浔品味装饰

生活不全是艺术，也是一种感受。功能本身就可以成为美，设计师试图用线条、光线、意象等元素，勾勒一个超越时间的空间，充分传达"静、空、奢"的意境。重心不仅仅是室内的结构设计，更要兼顾软装、饰件搭配，让现代风格和复古元素在此碰撞出当下非常流行的轻奢主义，低调地享受生活带来的欢愉。

Life is not all about art, it is also a feeling. The function can also be beautiful. The designer used lines, light, image, and other elements to outline a space beyond time, and conveyed the artistic conception of "serene, emptiness and exuberant ". The interior structural works with the combination of soft adornments, the modern style and retro elements collide with each other, all together, give out a vibe of chic luxury, and a kind of calm pleasure.

柔软的天

项目名称 柔软的天
设　　计 华浔品味装饰 / 周应钦、伍丽
建筑面积 480 平方米
项目地址 广州
主要材料 地砖、护墙板、不锈钢、大理石
工程施工 华浔品味装饰

SOFT SKY

本案根据内外空间环境的特点抽象出一类特征，传达一种生活审美，体现人文精神内涵。

营造一个可以让人感受到的氛围和气息，在小空间中减而不少，在大空间中丰富不塞，使其舒朗流畅、收放自如，给人带来美的享受。

According to the characteristics of the internal and external environment, a type of characteristics is extracted to convey the life aesthetic and humanistic spirit.

The project creates an atmosphere that can be felt by people. For small spaces, the abundance is retained with less elements; and for large spaces, the rich elements are arranged smoothly without causing overcrowding, bringing a comfortable living space as well as an enjoyment of beauty.

DONGJIU GARDEN

东氿壹号

项目名称 东氿壹号
设　　计 华浔品味装饰 / 潘骏
建筑面积 310 平方米
项目地址 宜兴
主要材料 大理石、KD板、仿真壁炉、大板瓷砖、玻璃、墙纸
工程施工 华浔品味装饰

本案为现代风格，去繁从简，不一味讲究元素堆砌，以简单到极致为追求。从入户开始到客厅、餐厅、卧室、书房等空间，功能各有侧重，动静走线明确，能够满足多种生活方式的需求。人生之乐，莫如自适其适，少即是多的设计黄金法则是生活中一种必要的断舍离。设计师想用最纯粹的方式打开一扇窗，极简亦是极美。整个空间亲切可近，线条平直，呈现出通透明朗的宽敞效果，给人干净利落的视觉享受，营造出现世安稳的感觉。

The project features a modern style that removes the complicated elements to pursue extreme simplicity.The entrance, living room, dining room, bedrooms and study each area has its own functions and clear circulation to meet demands of various lifestyles.The pleasure of life is to enjoy one's own life leisurely and comfortably. The golden rule of "less is more" in design is a minimalism, to open a window in the purest way and to achieve an ultimate beauty.The whole space is intimate and the lines are straight, building a spacious and transparent environment that provides a clean visual attractiveness and creates a peaceful atmosphere.

静美雅墅

项目名称 静美雅墅
设　　计 华浔品味装饰 / 卓天
建筑面积 720 平方米
项目地址 顺德
主要材料 大理石瓷砖、阿玛尼壁纸、柚木实木、天然大理石
工程施工 华浔品味装饰

SERENITY GRACE VILLA

既隔又合的空间在线与光的营造下，宁静而深邃，从容而淡定，东方美学意韵融满整个空间，在体现东方浑厚文化积淀的同时又营造出空与静的特征。

传统中透着现代，现代中糅着古典的设计手法使空间整体清雅含蓄，端庄丰华。在造型上，以简单的直线条表现中式的古朴大方，同时运用金属质感勾勒、点缀出生动而雅致的软装色彩。全屋地面和墙身筒一大理石瓷砖简洁大气给以稳重底蕴。背景墙面装点硬包与阿玛尼壁纸提升的质感恰到好处。整个空间给人一种静美舒适的笃定，让空间与人的对话无处不在。

The space is both separated and connected. With the lines and lighting, it appearsserene, profound and calm.The oriental aesthetics fills the whole space, expressing the rich cultural heritage and at the same time creating an atmosphere of emptiness and serenity.

Tradition is found in the modern. Classical design techniques are employed to achieve the overall elegance. Plain straight linesdemonstrate the simplicity of Chinese style, while the complicate shapes are eliminated. At the same time, metal texture is used to outline the lively and elegant colors in soft decor. The whole floor and wall finish is made of marble tiles with a sedate feeling. The hard background wall finish and the wallpaper enhance the texture. The whole space exudes anair of calm and comfortand also provides us with an elegant and relaxing quietness. Among this clean atmosphere, the dialogue between space and human is everywhere.

心地自清幽
A PEACEFUL HEART AT A PEACEFUL RESIDENCE

项目名称	万科水晶城
设　　计	华浔品味装饰 / 何志潮
建筑面积	810 平方米
项目地址	佛山
主要材料	大理石、实木地板、钨钢、艺术墙布、艺术涂料
工程施工	华浔品味装饰

本案是位于都市当中的别墅，业主言语中透露着大隐于市的情怀，对宁静生活的渴望，希望与自然为伍的心表。所以在设计中，设计师的思路是如何给业主营造一个独立的、不受干扰的宁静私人空间，于是就有了一个下沉式花园的设计，为了保证负一楼的通风采光，设计师在房子的一前一后各做了一个下沉式花园，然后负一层的各种生活功能都围绕着下沉式花园来布置，一改地下室那种密闭、黑暗、沉闷的感觉，反而使地下室变得通风采光、宁静清幽，与世无争。

The project is a villa in the city, the owner reveals his feelings of being hidden in the city and his desire for a peaceful life, closer to the nature. Therefore, my idea in the design is how to create an independent and undisturbed private space for the owners. So I made two sunken gardens, one before and one behind the house in order to ensure the ventilation and lighting of the basement, then all living functions are arranged around the sunken garden, the design change the impression and feeling of closeness, darkness and dullness in the basement, on the contrary, it makes the basement become airy, bright, quiet and undisturbed.

重塑理想生活

THE IDEAL LIFE

项目名称　融创博爵堡
设　　计　华浔品味装饰 / 周玉玲
建筑面积　330 平方米
项目地址　重庆
主要材料　乳胶漆、大理石、定制家具
工程施工　华浔品味装饰

客厅挑高的设计，超大落地窗的加入，让空间通透明亮，浅灰色墙漆、简约的线条设计，呈现现代美式的稳重优雅，美式家居的融入，营造出舒适、浪漫的空间氛围。在保证空间华丽精致品味的同时，也体现了钟爱高品质浪漫的生活居住者的个性品位。摒弃了浮夸与繁琐，在保证空间实用性的前提下，展现多姿多彩的空间面貌。

电视墙面的精雕细琢，把美式的高贵优雅之处表达出来，用马赛克瓷砖精心铺贴墙面，搭配石材雕琢的图案，给人高贵独特的视觉体验。

餐厅中无论是造型别致的金属大吊灯，还是抽象艺术挂画、餐具等不起眼的"小细节"，都能在无形中提升空间质感，打破单一性，让空间成为艺术品般的存在。

The high ceiling of the living room, along with the extra large french Window, make the space airy and bright. The light gray wall and simple lines present a kind of modest elegance. American style furnitures create a cosy romantic atmosphere. Abandoned grandiosity and fussiness, on the premise of ensuring function of the space, a colorful luxuriant space is presented, reflecting the owner's personal character and good taste.

With carefully laid Mosaic tile and delicately carved patterns on stone tiles, the TV backdrop wall manifests the American-style luxury and gives a unique vision.

The chic metal droplight, along with the abstractionism paintings, and all the " inconspicuous details", such as the exquisite tablewares …constitude the unique texture of the space, and make the dining-room a piece of art.

EUROPEAN GLAMOUR WITH ORIENTAL CHARM

项目名称 万科红郡
设　　计 华浔品味装饰 / 吴微
建筑面积 250 平方米
项目地址 广州
主要材料 奥特曼米黄大理石、手绘墙布、橡木护墙板
工程施工 华浔品味装饰

欧风东韵

本案是欧陆"新古典"风格于当代中国的"在地"演绎。轴线的对称、恢宏的气势、奢华陈设，尽显法式生活的优雅、浪漫和高贵。国画、屏风、瓷器、博古架、手绘墙布等中式元素的应用，也凸显着东方人所特有的风致与雅趣。

This project is the interpretation of the "new classical" style from the European continent in contemporary China. The symmetry of the axis, the magnificence, and the luxurious furnishings display the elegance, romance of French life, while the application of Chinese elements such as Chinese painting, screen, porcelain, antique shelf, and hand-painted wall covering also highlights the unique style of the orientals.

设计师在满足生活基本功能的前提下，重新规划空间布局，开辟更多用于社交、娱乐、休闲的功能空间。同时，还对主体建筑进行了局部改造，将自然的风景、阳光、空气导入室内，将建筑自然而然地置于优美的天然环境中，实现人、建筑、大自然的生态统一。

于是，天光云影、春景夏花、欧风东韵，尽享繁华。

Therefore, under the premise of satisfying the basic functions, the designer redesigned the spatial layout and created more functional spaces for social, entertainment and leisure. At the same time, the housewas partially transformed to bring the natural scenery, sunshine and air into the interior, tactfully placing the house in a beautiful natural environment to realize the ecological unity of people, buildings and nature.

LEAGUE OF LEGENDS

项目名称 英雄联盟
设　　计 华浔品味装饰 / 梁文仁
建筑面积 190 平方米
项目地址 佛山
主要材料 抛光瓷砖、乳胶漆、定制衣柜、大理石
工程施工 华浔品味装饰

英雄联盟

整个设计力求尽可能的干净利落，对于灯光氛围的营造、空间的沟通与趣味性上下了不少功夫，同时让模型成为亮点。

纯净的白色作为底色，穿插不同明度的灰，丰富空间色彩层次感，在低彩度的基色中运用少许高纯度色来避免空间的沉郁。在一个偌大的公共空间里，通过地面抬高的处理、灯光的营造，还有悬空的隔断装饰柜作为分区，隔而不断，以此来营造空间的趣味性和层次感！我们希望空间的是好像电影的镜头那样，所到之处都是一个完整的画面！都说光是空间的灵魂，自然光与人造光在这个空间中，交相辉映，仿若一场大合奏。

The whole design strives to be as clean and neat as possible, and it has made a lot of effort on the light-ing atmosphere, the communication of different areas and fun of the space, and at the same time highlights the model figures.

Pure white is set as the background color, interspersed with different shades of gray, to enrich the layering of colors, and a little high-purity color is added in the low-saturation base color to avoid depressing mood. In the large public space, the lifted ground, the lighting and the hanging decorative partition cabinet are designed to partition yet not separate the areas, increasing the cheerfulness and sense of layering in the space. We planto build spaces like the frames of a movie, every area we see is a complete picture. It is said that lighting is the soul of a space, here the natural lighting and artificial lightinglend radiance to each other to perform a magnificent ensemble in this space.

居逸·一江光华

A RIVER OF BRILLIANCE

项目名称　居逸·一江光华
设　　计　华浔品味装饰
建筑面积　230 平方米
项目地址　福州
主要材料　爵士白大理石、黑钛拉丝不锈钢、科技木面板、镜面柜门
工程施工　华浔品味装饰

这是一个春天完成的家居空间，设计师希望用洗练自然的笔触把春江美景画入家中，将室内和室外空间结合起来，为业主营造一个时尚而轻松、轻奢而自然的家居空间。

入户抬头，正对茶室，茶室以四扇窄边框玻璃做推拉门，玻璃与鞋柜的镜面相映成趣，奠定了质感卓越、宽松时尚的空间气质。坐卧其间，和家人好友品茗谈笑，琴声悠悠、书茶芬芳，看窗外"江上春山远，山下暮云长"。

Theproject is a home space completed in spring. The designer hopes to use natural brushstrokes to paint the beautiful scenery of spring river in the home, to take full advantage of the location and connect the indoor and outdoor views, creating a stylish, relaxed and natural living environment for the owner.

The tea room facing the entrance is equipped with a glass sliding door made of four glasses and narrow frames. The glass and the mirror on shoe cabinet set each other off, creating a stylish temperament with excellent texture. Sitting in the room, chatting with family and friends, listening to the beautiful piano music, enjoying the tea fragrance, admiringthe landscape of spring river outside the window.

进入客厅，室内外景致的融合更显巧妙，有序而不刻意，灵动而不杂乱。大面积的储物空间在沙发墙上，由整木面板和装饰层架组成。柜门以原木色和哑黑色为主，上下之间又有大理石条间隔，靠近窗户的部分采光条件更优越，所以选用稳重的色调，离窗的一端以原木色为饰，从光线到视觉从容过渡，自然与质感在这里相遇。

茶几和单位沙发在色彩和造型上更富于变化。透过巨大的落地窗，蓝天白云映于桌面，随着光阴流淌，围着桌上的绿植，自然与时尚在这里达成融合。

Entering the living room, the indoor and outdoor scenery is integrated in a more subtly way, orderly yet not deliberate. There is a large storage space on the wall behind the couch, consisting of wood panels and decorative shelves. The cabinet doors are mainly in wood and matte black colors, with marble strips between. The lighting conditions near the windows is better, so a darker color is applied; the part away from the windows is decorated with wood color. The transition from lighting to colors, is where the nature and texture meets.

The tea table and the single couchhave varied colors and forms. The blue sky and white clouds are reflected on the tabletop through the huge floor-to-ceiling windows, floating around the green plants on the table. Nature and fashion are combined harmoniously here.

新强苑
XINQIANG GARDEN

项目名称 新强苑
设　　计 华浔品味装饰 / 刘秀扬
建筑面积 600 平方米
项目地址 大浪
主要材料 大理石瓷砖、深木色原木、皮革
工程施工 华浔品味装饰

本案业主对新家的设计强调"偏爱质感",追求细腻、精致、有品质的生活空间。为此,设计师运用大理石瓷砖、深木色原木及皮革等自然材质,营造出一种轻奢的居家氛围。

本案设计师大胆运用高饱和度的色彩,精雕细琢的家具造型,金属、花纹元素,将都市轻盈的气质与欧式新古典的华美浓厚糅合,打造具有时代感的家。

The owner of this project emphasizes the preference for "fine texture" in design, and pursues an exquisite and high-quality living space. Therefore, the designer uses natural materials such as marble tiles, dark wood and leather to create a luxurious home environment.

The designer boldly uses colors with high-saturation, finely crafted furniture, metal and pattern elements, to combine the urbanrelaxed temperament and European neo-classical beautyto build a modern and fashionable home.

缘园人家
MODERN ORIENTAL RESIDENCE

项目名称　缘园人家别墅
设　　计　华浔品味装饰 / 陈健
建筑面积　300 平方米
项目地址　靖江
主要材料　大理石、科技布、护墙板
工程施工　华浔品味装饰

本案坚持设计的人文理想，通过对现代与东方的理解，简取"现代东方"的设计手法，刷新现代人居生活的定义，描摹品质生活的理想范本，营造美好生活情境，构建出优渥、简雅、怡人的都会美居。

Adhereing to humanistic ideal, based on profound understanding the meaning of modern and eastern, this "modern Oriental" design refreshed the definition of modern living, realized the ideal of quality life, created fabulous scenes of live, built a simple elegant place, a pleasant residence.

阑——霖峰壹号

项目名称　阑——霖峰壹号
设　　计　华浔品味装饰 / 唐璇
建筑面积　200 平方米
项目地址　南宁
主要材料　大理石瓷砖、实木饰面板、钛金不锈钢、刺绣硬包、硅藻泥
工程施工　华浔品味装饰

LINFENG GARDEN

本案原建筑为两间套房，设计将其打通变成一间大平层。整体的布局讲究对称感，融进天地方圆，在不经意中便体现出东方的韵味。

本案巧妙地结合东方元素与现代设计手法，在硬装上用黑钛金线条装饰，软装则保留了具有东方特色的基本线条，以"蓝"为主线，勾画出天与地、山与水、鸟语花香的融合，给人以自然之美，同时又不失现代时尚的感觉。在这种快节奏以及喧嚣的生活里，通过这样的设计，让生活慢下来，享受回家这一刻，在这"灯火阑珊"处……

The project is originally two apartments, combined into a larger one. The overall layout pays attention to symmetry, and blendsin oriental charm subtly.

Ingeniously combined oriental elements and modern design techniques, the project is designed with black titanium lines in hard decor, and basic lines with oriental characteristicsin soft decor. The blue color is the main thread, to painta natural beauty of sky and earth, mountain and water, as well as bird's twitter and flower fragrance while retaining the modern fashion. In this fast-paced and clamorous life, the design helps us to slow down the pace and enjoy the momentsat home, under the dim light.

NOBLE PENINSULA GARDEN

香缇半岛

项目名称 香缇半岛
设　　计 华浔品味装饰 / 李进强
建筑面积 220 平方米
项目地址 无锡
主要材料 奥特曼米黄大理石、手绘墙布、橡木护墙板
工程施工 华浔品味装饰

整个空间以灰色系呈现，让人感觉到一种大地自然的舒服；采用大面积的留白，通过木饰面为生硬的空间增加了温度。生活需要做减法，设计也是如此。

客厅主景没有选择传统的电视背景墙做法，而是选用白色的石材做不规则的造型，将客厅和过道以及楼上中空空间做了一个连接，达到贯通一致、创新独立的理想效果。

The entire space is decorated with black, white and grey tones to provide a comfort of nature for the owner to relax and return to the original pure state at home. A large area is left white and wooden veneer are employed to inject some warmth for this space. Life needs simplicity and so does design.

The main scene of the living room in irregular shape is made of white stones. Instead of setting a traditional TV background wall, the living room, the aisle and the hollow space of the upper floor are connected to achieve an ideal effect of consistency and originality.

设计上尽量运用开阔的设计手法,将餐厅、客厅相连,扩大视觉感受,打造出满足个性化生活需求的流畅动线。黑白灰色系与木饰面贯穿空间,灯具由几何线条塑造型,选择黑色金属灯杆迎合整体空间,增加美感度。

纵横空间,生活就在这里回响,除了延续原有整木的灰色系,家具的线条简约利落,还有灯具搭配手作的黑色铜质吊灯所碰撞的视觉变化,如同嬉皮般,追求那份自律性的自由,不在乎别人的眼光,勇敢地活出自己。

The designer tries to expand the visual space by linking the dining room and living room, creating a smooth traffic pattern. The black, white, grey tones and the wooden veneer runs through the entire space, and the luminaire formed by geometric lines has black metal poles that cater to the overall design.

Life is echoing here throughout the space. This area continues the original grey tone of wood, and the lines of the furniture are simple and orderly.The visual changes caused by the hand-made black copper chandeliers are just like hippiespursuing the self-disciplined freedom; they do not care aboutwhat other people think of them and live bravely to be their true selves.

著美·印记

BEAUTY · IMPRINT

项目名称　天湖郦都
设　　计　华浔品味装饰 / 何志潮
建筑面积　265 平方米
项目地址　佛山
主要材料　大理石砖、大理石、硬包、艺术地板、艺术涂料
工程施工　华浔品味装饰

业主是一对热爱生活，喜爱艺术品夫妻，他们对设计高度尊重和配合，才有了现在的高质量落地。玄关的四幅挂画都是从日本艺术家定制回来的画作；玄关、厨房、踏步位的彩砖都是专门定制的手工彩砖；墙面的艺术涂料都是从深圳找的意大利进口涂料做的，质感非常好。置身其中除了有生活的惬意，还有一丝丝艺术的气息，艺术源于生活，高于生活，更应融于生活。

The owner of the house is a couple who love life, admire art and respect the design. That is part of the reason of what the high-quality house is like now. All of the four paintings hung along the hallway are specially drawn by Japanese artists. All the bricks in the hallway, kitchen and the steps are custom-made manual colored bricks.
All the paint on the wall is from Italy and it has high quality.In addition to the comfort of life, there is a trace of art. Art should come from life, beyond life, and with life.

探索自由奥秘
THE FREE SPIRIT

项目名称	江山樾
设　　计	华浔品味装饰 / 周玉玲
建筑面积	180 平方米
项目地址	重庆
主要材料	中式家具、护墙板、乳胶漆
工程施工	华浔品味装饰

屋主一家5口人，有着不同的审美需求，因此设计上在满足现代实用风格的主基调上，去尊重彼此对方的喜好，在风格定义上没有太明显的界限，增加了一些美式轻奢的元素，整体效果明亮而清新，表达对生活本身的理解和对美好生活的无限向往。

我们尝试以清新淡雅亮的色彩与臻选的材料碰撞出眼前一亮的空间调性，素雅的色调邂逅明快的灯光，个性大方、自然沉稳，让每一个生活场景生动而鲜活。

All the five family members of this household have different aesthetic tendencies. We respect each member's preferences, and integrated everything in a modern functional framework. There is no obvious boundary separating one space from another. Some casual-luxury American elements make the whole place fresh and bright, expressing an upbeat expection for a better life.

We used fresh and bright colors to light up the space, together with carefully selected materials and amplitude lightings, we created a bright, natual and calming space for each individual and every scenes of life.

THE BLANK SPACE

留白

项目名称	复地源墅
设　　计	华浔品味装饰 / 黄江林
建筑面积	450 平方米
项目地址	无锡
主要材料	硬包、手绘绢画、黑钛钢、定制衣柜
工程施工	华浔品味装饰

水墨留白，虚实相生。

惜墨如金，计白当黑。

寥寥数笔丹青，

于方寸之地勾勒天地，

于无画处凝眸成妙境。

要说这现代中式，与传统中式相比，它用色更为大胆，形式更加活泼，结构也不讲究对称，给人感觉简洁舒适。为了保留中式的元素，设计师将具有传统风情的泼墨画广泛融入到客厅和卧室设计中。

Compared with traditional Chinese style, the modern Chinese style is bolder in selection of color and more lively in forms; and it does not require symmetry in structure, providing a feeling of simplicity and comfort. In order to preserve the elements of Chinese style, the designer integrates traditional splash-inkpaintings into the design of living room and bedrooms.

客厅：优雅的白色，自然的木纹，写意的水墨，干炼的线条，并运用现代块面状的设计手法，极力营造一种中国式的安静、优雅之美。
客厅沙发背景墙采用了一整幅的泼墨画硬包，留白之处让人可以有"此时无声胜有声"的想象空间。此外，电视背景墙选用一整幅爵士白大理石，与泼墨画交相辉映。

Living room: Elegant white color, natural wood grain, freehand ink, neat lines and the use of modern block-shaped design techniques are combined in this space to create a tranquil and elegant space of Chinese flavor. Sometimes we do not need too many materials to package our home, what we really need is a natural space to make ourselves feel happy.

The blank area of a painting can reflect the master's far-reaching thoughts and the high level of the design. The background wall finish behind the couch is a splash-ink painting, and the blank space on it leaves room for people's imagination. Additionally, the TV background wall uses a whole piece of jazz white marble t match with this painting.

星洲湾别墅
XINGZHOUWAN VILLA

项目名称 星洲湾别墅
设　　计 华浔品味装饰 / 刘文滨
建筑面积 680 平方米
项目地址 赣州
主要材料 大理石、板墙、壁纸、艺术涂料
工程施工 华浔品味装饰

本案在造型设计上化繁为简，精致的木雕点缀，保留传统中式风格含蓄秀美的精髓之外，重视文化意蕴，将传统中式进行时尚蜕变。家具陈设，讲究对称，每件家具都是有生命的，摆放在该空间，它们就决定该空间的气质，更彰显主人的品味与尊贵。

The projectsimplifies the complicated elementsby applying exquisite wood carving embellishment, reserving the essence of subtle beauty while also paying attention to cultural implications, transforming traditional Chinese style into a fashionable design. The furniture arrangement highlights on symmetry. Each piece of furniture has its own personality; they determine the temperament of the space where they are placed, and at the same time displays the owner's fine taste and status.

裕顺雅苑
YUSHUN GARDEN

项目名称　裕顺雅苑复式
设　　计　华浔品味装饰 / 庄佳佳
建筑面积　350 平方米
项目地址　南京
主要材料　涂料、皮革、仿古大理石、抛光砖
工程施工　华浔品味装饰

无论在时装界还是在家居界，黑白绝对是人人"趋之若鹜"的经典搭配。
它优雅得让人沉醉，简单得略显"鬼魅"，并以安静沉稳的力量释放出简约大气之美。
为了营造出简单却不失韵味的居住空间，设计师从点、线、面勾勒轮廓，从纯粹处填补颜色，巧妙的光影解构令整个场域回荡出无声的冲击。

The black and white combination is definitely a classic that everyone is pursuing, whether in fashion industry or in home designfield.

It is elegant and intoxicating, simple and slightly "ghostly", releasinga concise and sedate beauty with its calm power.

In order to create a simple yet appealing living space, the designer sketches the outline from points, lines and facades, and fills color in the pure places. The whole space echoesa silent impact with the help of the ingenious light and shadow deconstruction.

155

YUNXING ROAMING WONDERLAND

云星公园大观

项目名称	云星公园大观
设　　计	华浔品味装饰 / 刘文滨
建筑面积	272 平方米
项目地址	赣州
主要材料	仿古地砖、实木地板、墙纸墙布
工程施工	华浔品味装饰

褪去奢华，质朴拙雅的原味，这里有心灵对家的眷顾感……

从入户玄关的彩色仿古砖开始，空间都那么的自然风。家具陈设以古朴拙雅的美式风格为主。客厅木质墙板壁炉造型嵌入着电壁炉、宽松舒适的皮、布沙发，以及细腻柔美的提花地毯，加上电视背景墙仿古砖……美式乡村风格自然地流露在细微的每一处。

Taking off luxury, the original simple and elegant flavor is presented to us; here you can feel an affection for home…

Starting from the colorful antique tiles at the entrance, the entire space is exceptionally natural. The furnishings are mainlyin simple and elegant American style. Thewooden wall panel in the living room embellished with an electric fireplace, the large comfortable leather and fabric couches, the delicate jacquard carpet, plus the antique bricks on the TV background wall,are naturally revealing the American country style in every detail.

餐厅拱型造型墙与客厅、吧台区拱型门廊造型，既区分空间，又联系着空间。整体空间没有做过多的修饰和约束，不经意间却成就了另一种休闲式的浪漫，造就了整个空间自在、随意。

The arched shape of the dining room and the arched porch in the living room and bar area help to partition the spaces while letting them stay connected. There are not many decorations and constraints in the project, which inadvertently generates an air of leisure romance, creating a free and casual atmosphere.

南航——海岸华墅

COSTAL VILLA

项目名称	南航——海岸华墅
设　　计	华浔品味装饰 / 柯于锋
建筑面积	500 平方米
项目地址	海口
主要材料	木质家具、雅士白大理石、沙比利饰面、仿古砖、防腐木
工程施工	华浔品味装饰

整个空间都弥漫着一种对艺术充满热爱的生活环境。沉稳高雅的整体色调，装饰性的墙面，精致美好寓意的造型，以及极具感染力的艺术品都蓄意带来一种奇特而有趣的空间氛围。

客厅里深色木制家具，搭配温润色泽的皮质沙发，在明暗之间，形成了平衡的中性色调，沉稳精致又不失趣味，高贵精美又不失生活气息。顶上原是两根相交的梁，而设计师利用他多年经验将它做成了九宫格形状，既弱化了梁的突兀，又起到了装饰的效果。

The space is a living environment full of passion for art. The calm and elegant tone, decorative walls, exquisite and beautiful shapes, and intriguing works of art, all are deliberately bringingan unusual and fascinating atmosphere.

The dark wood furniture in the living room is joined by a warm leather couch, between the light and dark, will form a balanced neutral tone, calm, exquisite and yet interesting with a flavor of life. There was originally two intersecting beams on the ceiling, with his rich experience, the designer decorated them into a nine-square form, which weakened the abruptness of the beams and functioned as an embellished at the same time.

九里峰山

JIULI FENGSHAN

项目名称 九里峰山
设　　计 华浔品味装饰 / 洪文龙
建筑面积 800 平方米
项目地址 赣州
主要材料 原木，铜艺，绸缎，布艺，新西兰羊毛，皮革，水晶，大理石
工程施工 华浔品味装饰

设计师将"小隐于野，大隐于市"的生活愿景注入到设计中，将东方韵致与现代美学融合为一体，演绎出传统的人文情怀。别墅布局分为三层，负一层设置了茶室，书房，影音室和健身房；一层是客厅与餐厅和长辈房；二层是儿童房及起居室；三层则是主卧及小女儿房。游走其中，便能领悟现代别墅应有的尺度与气度。

Citing the core value of "the "lesser hermit" lives in seclusion in the country, the "greater hermit" does so in the city", the designer presents a home with a combination of oriental charm and modern aesthetics. The three floor house boasts a remarkable contemporary lifestyle. The basement features a tea room, a reading room, a media room and a gym; the living room, kitchen, dining room and grandparents' bedroom are on the first floor; the second floor contains a children's room and a family room; the third floor comes the master bedroom and the little daughter's room.

本案定位新中式风格，整个空间贯穿隐喻的表现手法来渲染空灵写意的东方意境。利用富有变化的色彩与软装陈设。讲求意境，拒绝铺陈，达到东方精神境界的追求，让静谧自然，清幽静默如一股清泉般静静的流淌在整个空间。

The Neo-Chinese style incorporates an ethereal and dreamy oriental factor that is further heightened with flexible color and soft furnishing setting, layering an underlying mood of tranquil serenity. With simple but delicate arrangement, the home is dominated by lingering oriental charms that flow like a breath of fresh air.

远洲墅

YUANZHOU VILLA

项目名称 远洲墅
设　　计 华浔品味装饰 / 卢光
建筑面积 500 平方米
项目地址 黄岩
主要材料 仿古砖、硬包、不锈钢、仿真壁炉
工程施工 华浔品味装饰

本案不刻意塑造强烈的造型风格，以低调内敛闲室的奶灰色系为主。深灰色哑光地面、墙面奶灰色平板扶墙，再植入黑色钢板立面装饰柜，呈现有气质而又不张扬的内敛格调。考虑空间与人之间的关系，本案更注重推崇"以人为本"、"环境融入设计，设计融入艺术"的设计精神。强调造型简约、尺度合宜、色调协调的装饰设计理念。

The project did not try to create a strong style, instead it is decorated with an understatedand introverted milky grey color.

The dark gray matte floor, milk gray wall panel and the black steel finish decorative cabinet presents an elegant andrestrained style.

In consideration of the relationship between space and human beings, this projectfocuses on building ahuman-oriented space which "integrates the environment into design and design into art". The design adheres to the concept ofsimple shapes, proper proportion and harmonious tones.

蓝色多瑙河

THE BLUE DANUBE

项目名称　龙湖别墅
设　　计　华浔品味装饰 / 冷婷婷
建筑面积　400 平方米
项目地址　烟台
主要材料　仿大理石砖、硅藻泥、涂料、大理石
工程施工　华浔品味装饰

女主人温和得体，对人友善极其尊重对方的付出，喜欢温馨舒适的环境，美式在这方面有极大的优势。男主人小心谨慎，性直执拗，对新生事物接受较慢，对传统文化还是存在着极大的兴趣，于是以西方古典方式处理空间结构，以东方文化植入西方空间结构之中，表达出东方的审美情趣。

The owner is a friendly lady who prefers a cozy environment, therefore the American style is chosen to meet her requirement. Her husband is more stubborn and casual in taking in new things, and has a great intrest in traditional culture. Hence the oriental culture is integrated into the western Classical spatial structure to express oriental aesthetics.

格局上,没有跟其他的户型一样将采光井加建起来。北边视野开阔,小区的主景观区域尽收眼底,所以将主入户加大,不但增大了收纳及换鞋的区域,还增加了二楼主卧外面的观景平台的面积,清晨醒来,推开玻璃门,沐浴在阳光和美景下。

白茹凝脂,蓝含月润,珍珠白沁就烟雨,孔雀蓝映著月光,莹透的素颜,朦胧了琴弦上一缕香,案子就取名为蓝色多瑙河。

In the layout, this project has a full view of main landscape of the community at the north side. Therefore the main entrance area is widened to increase the storage space and the viewing platform outside the master bedroom on the second floor. Pushing open the glass door, the owners can easily access to the sunshine and scenery every morning.

EMINENCE

项目名称　名门壹品别墅
设　　计　华浔品味装饰 / 胡戈
建筑面积　400 平方米
项目地址　扬州
主要材料　大理石、实木、硬包、墙布、不锈钢
工程施工　华浔品味装饰

名门壹品

于简洁大方间沉淀神髓，
在含蓄内敛中绽放凝练，
一如水与墨的交融，
无声无息，芳华俱现。
设计师在此融入中国文化底蕴的同时，注入创新时尚元素，达到挥洒自如、一气呵成的艺术境界。韵味不在于多，而在于精，力图在传递大气优雅的高品质生活方式同时，使居者切实地感受深厚绵长的东方古韵。

While incorporating the Chinese cultural heritage, the designer also injects innovative and fashionable elements to achieve a freely and smoothly flowing artistic state. The charm is not in the quantity of the elements, but in the essence; the intention is to convey the profound ancient oriental charm to the residents whileexpressing an elegant and high-quality lifestyle.

端庄丰华恢弘大气，一股独特的东方韵味，恰如其分地营造出成熟稳重，又不失舒适、雅致的空间。简洁的几何线条勾勒出完美的天花，浅薄浣纱的引入开阔视野，配以精致的艺术吊灯、全套古色的家具与水墨画地毯来提亮空间，巧妙地打造出室内的层次感，营造"别业居幽处，到来生隐心"的美妙氛围。

The living room is dignified and splendid with an unique oriental charm, properly creating a mature, sedate,and at the same time comfortable and elegant space.The simple geometric lines compose a perfect ceiling finish, and the thin gauze introduces an open view, adorned with exquisite art chandeliers, a full set of antique furniture and ink wash painting pattern carpet to brighten the spaceand build the layers of the interior delicately, creatinga poetic atmosphere of "a house situated in a quiet scenery that makes one feel like returning to pastoral life".

让岁月留香来日可期
FRAGRANT DAYS

项目名称　金茂墅
设　　计　华浔品味装饰 / 周玉玲
建筑面积　240 平方米
项目地址　重庆
主要材料　护墙板、中式家具、乳胶漆
工程施工　华浔品味装饰

房子共 4 层，根据屋主的设计需求，负二层是车库及健身活动区，负一层是棋牌茶室，一层是客餐厅，二层是卧室。设计师匠心独运地将古典韵味融入到家居每一个空间，使"新中式"更加实用、更富现代感。

The house has a total of four floors. Designed according to the requirements of the owner, the second floor undergrand is the garage and fitness area, the first floor undergrand is the game room for chess and card etc. The living room and the dining room for guests are on the first floor, and the bedrooms are on the second floor. The decorator ingenuity blend the classical charm in, and made the "new Chinese style" design more functional and chic.

在客厅的布局上，保留了传统的对称结构，选择淡雅温润的木色家具。改造后的客餐厅空间采光很棒，地砖选用细微的青花瓷色彩混搭雨林啡石材的边带，提升新中式色彩细节。沙发一侧是收藏柜，用来摆放屋主平时收藏的石头，中式元素与现代材质的巧妙结合，达到传统意韵、灵净素雅、上善若水的独到新中式之风。

The living room has a traditional symmetrical layout, furnitured with wood of soft colors. After re-construction, the guest dining room is pretty spacious and has marvelous natural lighting, on the floor tiles, chinese blue and white porcelain was mixed with some rain forest elements, giving more color details to this "new Chinese style" design. By the sofa is the cupboard for the owner's stone collections, Chinese elements and contemporary materials are cleverly combined, constructed a simple but elegant, new Chinese style masterpiece.

RESIDENTIAL SPACE

公寓住宅空间

2020 / 品味 TOP 100

184	滟澜公馆 GLISTENING WAVE MANSION	228	铜锣湾广场 CAUSEWAY BAY PLAZA
188	江与城 THE RIVER AND THE CITY	230	橙意浓 ORANGE
192	黑白之间，独爱那一抹橙 A TOUCH OF ORANGE IN THE MIDST OF BLACK AND WHITE	234	棕榈泉平层 PALM SPRINGS FLATS
196	四季如春 SPRING ALL YEAR ROUND	238	富贵红心 A DASH OF RED
200	凤凰汇·熙园 GALLERIA · XI GARDEN	242	旺庭公馆 WANGTING MANSION
204	新中式·韵 NEO-CHINESE STYLE · RHYTHM	246	童话·孔雀和天鹅的游戏 FAIRY TALE, THE GAME OF PEACOCK AND SWAN
206	归·家 RETURNING HOME	250	韵·律 RHYTHM
210	世纪嘉园 CENTURY GARDEN	254	恒利·江悦明珠 HENGLI · JIANGYUE MANSION
212	流动的溢彩 FLOWING COLORS	256	天马相城 TIANMA VILLAGE
216	华发荔湾荟 FLOURISH GARDEN	260	上府名园 THE HOUSE GARDEN
218	邂逅闲适浪漫 ENCOUNTERING LEISURE & ROMANCE	264	异 境 WONDERLAND
222	朝雨浥尘 RAIN AND DUST	268	融景城 FUSION CITY RESIDENTIAL AREA
224	舒·宁 COMFORT·SERENITY		

GLISTENING WAVE MANSION

滟澜公馆

项目名称 滟澜公馆
设　　计 华浔品味装饰 / 孙丹萍
建筑面积 220 平方米
项目地址 徐州
主要材料 原木、竹、藤、麻、艺术漆、硅藻泥
工程施工 华浔品味装饰

本案设计为日式风格，整个空间干净，纯粹，有禅意，回归自然。色彩方面多以白色和原木色为主。大多以典雅的色调为主，色彩多偏重于原木色，以及竹、藤、麻和其他天然材料颜色，形成朴素的自然风格。造型方面多为直线条，大面积的留白形成空间流动与分隔，流动则为一室，分隔则分几个功能空间，装饰和点缀较少，简洁大方，有一种置身于大自然的感觉。空间中总能让人静静地思考，禅意无穷，也体现房屋主人淡泊宁静、清新脱俗的生活态度。

The project is designed in Japanese style. The whole space is clean, pure and filled withZen. The colors are mostly white and wood, emphasizing on the color of the wood and bamboo, rattan, hemp and other natural materials to form a simple natural style. The forms in this spaceare mainlystraight lines, the white spaces are applied as both a connection and partition between different areas. When partitioned,the space is divided into several functional spaces. Without too many ornaments and embellishments, the spaceprovides a feeling of being in nature. This space allows people to ponder quietly which reflects the peaceful state of mind of the owner.

一层客餐厅墙面装修以"白"为主，没有太多的色调拼接，用最干净的清淡色彩来表现出一种极简的自然主义。墙面装修材料的选择，也贴近一种自然之感。玄关处表露出水泥表面，着意显示素材的本来面目，加以精密的打磨，表现出素材的独特肌理。家居以简约为主，整体家居中强调的是自然色彩的沉静和造型线条的简洁，家具低矮且不多，给人以宽敞明亮的感觉。

The wall finishing of thedining room in the first floor is mainly in white color. There is not many colors in this area. The most clean and plain color is used to express a minimalist natural style. The cement at the entranceway is revealed todisplay the original features of the material and the unique texture. The whole home space highlights the calmness of the natural colors and the simplicity of the forms.

江与城
THE RIVER AND THE CITY

项目名称 江与城
设　　计 华浔品味装饰 / 刘云
建筑面积 312 平方米
项目地址 重庆
主要材料 原木家具、木地板、艺术漆
工程施工 华浔品味装饰

认识苏姐四年，缘起孩子，十几个陪读家长中，安静看书，偶尔抬头看看娃的她给我留下极深印象。以后的日子，每天放学后的水果分享会，娃儿们的英语小课堂，以及我们倒腾的小事业，虽无疾而终也属趣事一件。都是散漫而热爱生活的人，所以当苏姐决定装修新居时，我们之间对于家的探讨更深一步。

I got to know Mrs Su because of our children. Among dozens of parents accompanying their children, I am deeply impressed by her. She was always reading quietly and looking up at her kid occasionally. From then on, we had a lot of fun memories together, such as fruit parties, after-school english class, and our little business. Since we are both life-loving, casual person, we had a more profound discussion about home, when Mrs Su decided to decorate her new house.

最平淡的居家生活就是美学的游走及传播，一花一木、一器一物，重视四季的变幻，感知生活中的美。做有温度的、有生活气质的设计。春则觉醒而欢悦、夏则在小憩中聆听蝉的欢鸣、秋则悲掉落叶、冬则围炉夜话雪中寻诗，生活中实践美学的她也在着力创造人性化，诗意化的空间。

Beauty walks in the most common daily life, flows in each flower, each tree, each item, each season. A good design should be warm, should perceive and celebrate the beauty lies in the daily life.

In spring, she awakens and rejoices; in summer, she listens to the cicadas' chirping during a nap; in autumn, she mourns the leaves'fall; in winter, she reads poems around the fireplace or in the snow. she is always exerting herself to create a humanized and poetic space.

A TOUCH OF ORANGE IN THE MIDST OF BLACK AND WHITE

黑白之间，独爱那一抹橙

项目名称	海岸城郦园
设　　计	华浔品味装饰
建筑面积	170 平方米
项目地址	无锡
主要材料	米灰墙纸、黑色墙饰、木地板
工程施工	华浔品味装饰

本案在空间布局上进行重组，将原始狭长的过道划分成三块，自然形成独立门厅的同时，又增加了一个储藏室。公共活动区域和休息区域完全分离，增加了卧室空间的私密性。利用北面阳台的空间将厨房外移，做成中西式兼具的厨房，使之更加符合现代人的生活方式。设计上采用结合点、线、面的处理方式，黑色的墙饰及地板搭配大块面米灰色的墙纸，充满活力的橙色在整个空间的贯穿和点缀，让空间既沉稳大气又充满趣味性！

The project layout was redesigned by dividing the original narrow aisle into three areas, naturally forming an independent entranceway and an extra storage room. The public area and the bedroom area are completely separated, which increased the privacy of the bedrooms. The kitchen was moved outward to the original space of the north balcony, to create a kitchenthat is able to serve both Chinese and Western cooking, catering to the demands of modern lifestyle.
The design techniques of point, line and plane treatment are applied. The black wall finish and the floor are matched with the rice gray wallpaper. And a vibrant touch of orange brightenedthe whole space, infusing a pleasure to this calm environment.

四季如春
SPRING ALL YEAR ROUND

项目名称 四季如春
设　　计 华浔品味装饰
建筑面积 170平方米
项目地址 无锡
主要材料 护墙板、涂料、皮革、新中源微晶砖
工程施工 华浔品味装饰

本案在空间布局上进行调整，把原始零碎的空间重新组合，让各个空间变得更加整体，使之更符合现代人的居住生活方式。雪松绿作为大面积的背景色，让空间看起来生机盎然，饱含着一股春天的气息，营造出一种优雅轻松的氛围，再点缀一抹跳脱的爱马仕橙，整个空间产生丰富的视觉层次，就像是身着华服的贵妇人，举手投足间的雍容姿态，折服着每一个人。所有的家具、饰品、墙纸的搭配，都是静心挑选的各自互补的色系，让空间变得协调而鲜明，更充满着一份尊贵的气势。

The spatial layout was adjusted to combine the original fragmental spaces, so that each space was reorganized to cater to the demands ofmodern lifestyle. As the background color, the cedar green gives a vibrant look and fill in the space with an air of spring, creating an elegant and relaxed atmosphere. A touch of Hermès orange is added to create a rich visual layering, just like a gorgeously dressed lady who impressed everyone with her dignified bearing. All the furniture, accessories, and wallpapers are carefully selected to match each other in the tones, forming a harmonious, bright andstately space.

凤凰汇·熙园

GALLERIA · XI GARDEN

项目名称 凤凰汇·熙园
设　　计 华浔品味装饰 / 井峰
建筑面积 220 平方米
项目地址 盐城
主要材料 薄板、PU、不锈钢
工程施工 华浔品味装饰

本案业主对平面空间规划创意相当重视，经过沟通后，设计师把原有两个卫生间位置改为楼梯间，整个空间的动线更加舒适性，也使楼梯间更加宽阔。原有的小厨房和一间小卧室合并为中西餐厨房。业主对后期的风格希望尽量简洁，提倡空间简洁明了，同时略带一丝韵味。设计师打破了传统新中式风格，选用了线条感清晰的家具，同时对墙面的线条装饰进行呼应。后期软装饰品的搭配更是选择了一些具有现代金属感的装饰品和一些亮色的布艺点缀，让整个空间更加生动。

The owner pays high attention to creativity in the planning the layout of this project. After discussion, we moved two bathrooms to make space for a wider staircase and to create a better interior circulation.The original space of a kitchenette and a small bedroom are combined into a Chinese-Western kitchen. The owner hopes the space to be as concise as possible, with a bit of its own charm. Therefore, the designer broke the conventions of Chinese style and adopted some furniture with clear lines that echoes to the line decoration on the wall. In soft decor, the project even chose ornaments with modern metallic feel, and some bright color fabrics toconstruct a morevivid space.

新中式·韵
NEO-CHINESE STYLE · RHYTHM

项目名称 新中式·韵
设　　计 华浔品味装饰
建筑面积 410 平方米
项目地址 佛山
主要材料 大理石、硬包、中式木格、黑色不锈钢装饰条
工程施工 华浔品味装饰

业主之前一套房子装修因其主观意识太强，设计师都跟着业主的思想走，最终装修效果不尽人意没达到预期。在设计之前经过了解后，本案例完全按照设计师的理念，经过沟通最终建议确定新中式风格。通过对传统文化的认识，将现代元素和传统元素结合在一起，将传统的元素融合在现代的设计中，让人既体会到现代设计的科技性，又体会到传统的中国韵味。

After communicating with the owner, the designer decided to apply Chinese style in this project. The traditional elements are combined with modern elements to infuse traditional flavor into modern design. In this way, the owner can experience the convenience brought by technologies in modern design and at the same time appreciate the traditional Chinese charm.

归·家

RETURNING HOME

项目名称 归·家
设　　计 华浔品味装饰 / 何俊杰
建筑面积 105 平方米
项目地址 福州
主要材料 诺贝尔瓷砖、书香门第地板、西顿照明、芬琳进口漆、爱玛仕家具
工程施工 华浔品味装饰

目前是三口之家，考虑到父母偶尔会来住，今后又有考虑再生一个，所以房间三间怕后期不够住，于是女儿房那间选用了上下床，以便后期两个小孩都可以住。在空间规划中相对来说没有太大的改动。在选材方面，因为小孩比较小，更多地考虑环保，实用为主。在配色上，选用了蓝绿色，使得空间充满活力。卧室选用蓝色作为背景色，整个空间比较浪漫。

The owner is currently living with his wife and daughter, and his parents comesto stay occasionally. The couple is also considering to have a second child this year, three bedrooms will not be enough to accommodate six persons, therefore a bunk bed is installed in the daughter's bedroom. There is relatively little change in spatial layout for cost reasons. In terms of material selection, environmental protection and practicality are the primary concerns as there will be two young children. The blue-green color is applied to create a vibrant space and the blue color is used as the background in the bedrooms to achieve a romantic ambience.

后期软装主要体现两大概念，一个是在沙发后面的抽象时钟，代表着时光、岁月，珍惜当下拥有的；另一个则是大海沙滩脚踏的脚印那幅画，则是告诉自己与家人，做人做事要心胸宽广，要脚踏实地，一步一个脚印，这样的路才能走得长久。整个空间给人一种干净的感觉，让我们每天上班后想回家安安静静地休息，享受家的温暖。

There are two mainconcepts in the soft decorations, one is the abstract clock behind the couch, which represents the time and reminds one to cherish what you have in your life; and the other is painting of footprints on the beach, to tell himself and the family to be broad-minded and down-to-earth. One step at a time and it will eventually take you there. In conclusion, thespace provides a clean environment for the family to rest quietly after returning from their daily work and to enjoy the warmness of home.

世纪嘉园

CENTURY GARDEN

项目名称	世纪嘉园
设　　计	华浔品味装饰 / 钟博
建筑面积	140 平方米
项目地址	赣州
主要材料	金意陶仿古砖、硅藻泥、整木定制、圣象地板
工程施工	华浔品味装饰

优雅于里、简洁于形、无须造作、自由自在，一位年轻貌美女士追求这样优雅、自由、理想的品质空间。在设计思路上，设计师利用圆弧的造形、拱门形的门廊，用纯色点缀空间相辅相成，相得益彰。地中海风格具有鲜明的特色，家具利用纯本色装饰，线条简单修边浑圆的本质家具。地面铺贴赤陶点缀的马赛克，彰显安宁静谧、自由奔放、色彩鲜明的装饰，带你走进慢节奏尽享浪漫的生活。

Elegant in the inside, simple in shape, with free expressions, without any pretentious element; the project is the ideal elegant space that a young and beautiful lady pursues. Arc shape is usedat the arched porch and solid colors are employed to adorn the space.The Mediterranean style has distinctive features. The furniture is adorned with pure natural colors, and the lines are simple and trimmed. The floor is covered with terracotta-adorned mosaics, to highlight the tranquil, unrestrained and colorful decorations, bringing you into a slow-paced romantic life.

流动的溢彩
FLOWING COLORS

项目名称 流动的溢彩
设　　计 华浔品味装饰 / 何俊杰
建筑面积 150 平方米
项目地址 福州
主要材料 唯美 LD 瓷砖、得高进口地板、玛格衣柜、西顿照明、本杰明进口漆、巴里巴特家具
工程施工 华浔品味装饰

本案业主为三口之家，女儿读高中，父母均为国有企业员工。业主喜欢美式轻奢风格，对空间布局有自己的想法。在色彩方面，选用暖色调，在后期软装搭配中带点撞色，使空间显得有活力，充满色彩感。

The owner of this project is a family of three, who prefersAmerican style with accessible luxury, and has original ideas about the layout design which was finalized as threebedrooms, one living room, one dining room, two bathrooms and one study. They favor warm tones andthere is also a bit color contrast in soft decorations, creating a vibrant andcolorful space.

本案通过沟通，设计师在设计过程中先从空间规划入手，在合理的布局情况下，通过材质的运用、色彩的搭配、灯光运用，使得空间充满设计感。设计主要体现去繁就简的生活追求，纯正色彩的运用非常能表达出这种感觉，线条感明快的家具犹能体现出一种简练美，贴合快节奏时代同时，更可以传达出现代人对高品质生活的追求，使整个空间洋溢出一股美式轻奢的气息。

After communication, the designer starts with spatial planning in the process. With a rational layout settled, a creative space is constructed through the selection of materials, color matchingand lighting. The main concept is the pursuit of simple life, which is fully expressed by the application of pure colors. The furniture with bright lines presents a concise beauty that caters to this fast-paced era and at the same time conveys the modern pursuit of high-quality life, filling the entire space will a flow of American-style luxury.

华发荔湾荟

FLOURISH GARDEN

项目名称 华发荔湾荟
设　　计 华浔品味装饰 / 袁红霞
建筑面积 140 平方米
项目地址 广州
主要材料 IMOLA砖、德合家进口木地板、舒尔茨进口墙漆、哑光黑钢、钢化玻璃、harbor house家具
工程施工 华浔品味装饰

本案业主是一位帅气、温文儒雅的金融工作者，由于平时工作繁忙，没有太多休假的机会。为舒缓业主快节奏的生活步调，设计师特别将空间塑造成一个机能与休闲并重的居家空间。

整个客厅以白墙为底，用艺术性的简单线条造型及家饰，以及对比的色彩效果，营造了一个视觉丰富的空间世界。极具设计感的装饰、雕塑、挂画，看似随意的添置，为空间增添不凡的艺术气息，整体空间理性中不乏精致，塑造业主不凡的品味与形象。

The house owner is a handsome gentle financial worker, because of the busy work schedule, he has very little time for vacations. In order to give the owner a relief from the fast pace of life, the decorator made his home a relaxing place as well as a functional house.

The sitting room has simple white walls, and decorated with clean lines, artsy furnishings, contrast colors, resulting a luxuriant looking space. Finely designed ornaments, sculptures and paintings create an extraordinary artistic atmosphere, while the overall space gives a rational and refined impression, conforming to the owner's extraordinary taste and image.

ENCOUNTERING LEISURE & ROMANCE

邂逅闲适浪漫

项目名称 泰禾红誉
设　　计 华浔品味装饰 / 王子来
建筑面积 146 平方米
项目地址 福州
主要材料 大理石护墙板、定制酒柜
工程施工 华浔品味装饰

每个人都渴望，
栖身于世外桃源般的安和，
因此每一座城市，
都有一片静谧之地，
但长乐这座城市本身，
便就不慌不忙，
不偏不倚，
是人们落定心灵的归宿！
入户玄关通过黑钛搭配大理石护墙板，给人坚硬的力量感但又不失细腻。

The entrance hall uses black titanium and marble asthe wall covering, giving a feeling of strength yet with delicacy in it.

客厅地面选用暖色系，配备的沙发起到了调和色彩的作用。让人过目不忘的孔雀蓝沙发，温柔优雅的单人沙发，让这个空间立刻鲜活起来。电视背景墙颜色上和整体空间保持相对的统一性，设计错落有序，增添了整个空间的层次感。

开放式的餐厅简洁大气，黑白相间的皮质座椅，黑色的大理石餐桌，增添了一丝华丽。餐厅独特的长吊灯处理，增加就餐时的趣味。

The living room is designed with a warm tone floor and a couch is set to blend the colors. The striking peacock blue couchand the gentle and elegant single sofa chair enliven this space.

The color of the TV background wall is in accordance with the overall space, and the decorations areorderly scattered, adding a spatial hierarchy in the space.

The open dining room is simple and elegant, while the black and white leather seats and black marble dining table adds a touch of gorgeousness. The unique long chandeliers increases the fun of dining.

朝雨浥尘

RAIN AND DUST

项目名称 朝雨浥尘
设　　计 华浔品味装饰 / 侯国富
建筑面积 140 平方米
项目地址 佛山
主要材料 灰白色瓷砖、木纹砖、雅士白大理石、烤漆饰面、不锈钢、金属板、巴宝莉皮
工程施工 华浔品味装饰

渭城朝雨浥轻尘，客舍青青柳色新。轻松，是空间改造要达到的目的。铅华洗尽，剩下的是直面内心的独白。我们尽力追求简单纯净的空间，用简单的几何构图，让空间和光影更好地交织互动，点缀几许巴宝莉的真皮和原木，让空间多几许朝雨润过后的自然清新气息。

Relaxation is the goal of this house renovation. All that is left is the monologue of the heart. We try our best to create a simple and pure space with using simple geometric composition to make space and light interact better. Embellished with Burberry's leather and logs, the space seems filled with a fresh air after the morning rain.

舒·宁

COMFORT · SERENITY

项目名称 舒·宁
设　　计 华浔品味装饰 / 何俊杰
建筑面积 120 平方米
项目地址 福州
主要材料 诺贝尔瓷砖、书香门第地板、西顿照明、芬琳进口漆、康宝 VV 家具
工程施工 华浔品味装饰

本案业主比较时尚，在做方案前期做了大量的沟通，我们基本知道她所想要的后，决定以舒适、宁静这个设计理念来贯通整个设计。在空间规划中相对来说没有太大的改动，因为原有户型基本布局还可以。在选材方面，因为女业主比较喜欢大理石，在背景中适当地使用一些大理石来衬托。其他材料主要以环保、实用为主。在配色上，选用了黑白灰，为主色调使得空间充满时尚感。在后期软装上，主要以蓝色作为撞色，提亮空间色彩感，使得后期业主在生活过程中，在家中享受宁静优雅，舒适的回归。

The owner of this project is a fashionable female. Based on her demands, the designer decides to set comfort and serenityas the design concept. There is relatively little change in the spatial layout. As the owner prefers marble, some marbles are used in background finish while other materials used are all environmental friendly and practical. The spaces are all fully utilized. In color composition, black, white and grey are applied as the main tones to paint afashionable space. In the soft decorations, the blue color is employed to createcolor contrast to brighten the space, allowing the owner to enjoy a tranquil, elegant and comfortable family life here.

铜锣湾广场
CAUSEWAY BAY PLAZA

项目名称　铜锣湾广场
设　　计　华浔品味装饰 / 胡京军
建筑面积　100 平方米
项目地址　南昌
主要材料　大理石、软包、乳胶漆、定制橱柜
工程施工　华浔品味装饰

本案采用的是混搭风格,设计师将其打造出鲸游夏夜般的美妙且富有生命力的空间。混搭风格没有固定的主色调,为了让空间不显得繁琐复杂,设计师确定了三个基本的色调,灰蓝、灰绿、橙色为主,金色、黄色作点缀。室内的每个空间都有其特定用途,设计师使每个空间之间既能随意转换,又具有独立性。

客厅空间以沉稳的米白色为基本色,金色、灰蓝色、灰绿色、橙色点缀局部,格调高雅富有质感,同时也增添独特的空间层次感。

This project adopts a fusion of different styles. The designer creates a wonderful and lively space. There is not a fixed main tone in this style, in order to avoid complication in appearance, the designer sets grey blue, grey green and orange as the main colors, while using gold and yellow as embellishment. Each room has its own function,and is convertible and independent at the same time.

Calm beige is the main tone in living room, adorned with golden, grey-blue, grey-green and orange in details, creating an elegant environment and adding a sense of layering in space.

橙意浓
ORANGE

项目名称 天奇城
设　　计 华浔品味装饰 / 孙正川
建筑面积 143 平方米
项目地址 无锡
主要材料 调色混油木作、爵士白大理石、进口艾瑞斯曼壁纸、仿玫瑰金拉丝不锈钢条
工程施工 华浔品味装饰

本案业主为企业高管，恩爱夫妻陪伴一对儿女。业主对设计效果、空间生活功能及造价重点关注，所以在施工用材上，使用比较常规和朴素的材料。设计手法上，始终运用简洁的造型，通过色彩的转换与材料自身质感，去表达空间之美。

This project provides design and construction services for a family. The loving couple are corporate executives and havetwo children. They pay high attention to the design,functions and costof the space. Therefore, more conventional and simple materials are applied. In design techniques, the project always uses simple shapesto express the beauty of this space through the chemical reaction generated by color conversion and the texture of the materials.

棕榈泉平层

PALM SPRINGS FLATS

项目名称	棕榈泉平层
设　　计	华浔品味装饰 / 刘云
建筑面积	160 平方米
项目地址	重庆
主要材料	饰面板、木地板、硅藻泥、实木
工程施工	华浔品味装饰

本案大面积采用中性色调，搭配暖调质感原木，呈现大气明朗视觉效果。客厅采用水墨感大幅挂画，地毯也一样，给空间渲染出一丝东方意蕴。 沙发整体以简单的深灰与白色、木色相互搭配，水墨感大幅挂画及地毯渲染出一丝东方意蕴，也让整体空间层次感更强。重点落在弧度饱满的两把单椅上，正面、背面、侧面造型皆完美。

一门之隔的书房伴着清风徐来，端坐案头亦可，卧榻一角更欢。卧室里柔软的床头伴君入眠，床品质感松软舒适。

The apartment presents bright and clear visual effect, with neutral tones used in most areas and warm tones in logs. There are large ink paintings and carpets in the living room which not only convey an oriental sense but also create a sense of layering. As for the sofa, the simple dark gray matches perfectly with white and wooden color. The highlight is two side chairs which have wonderful angles. And the shapes of front, back and side are all perfect.

In the study, you can sit by the desk or lie leisurely on the floor. Oh, do not forget the soft and comfortable beds in the bedroom.

富贵红心

A DASH OF RED

项目名称 中央御景
设　　计 华浔品味装饰 / 孙正川
建筑面积 163 平方米
项目地址 无锡
主要材料 染色柀、山水纹大理石、闷布、进口艾斯曼壁纸
工程施工 华浔品味装饰

芸芸山水间，道法从自然；
锦袖乾坤现，天地美不言。
本案采用中式传统隐喻的设计手法，以物咏志，以简代奢，从细节处理上去体现人性化，用色彩和陈设上的考究，去表达精致细腻的家居之美。

Amidst the mountains and rivers, the Tao models itself after nature;
Heaven and Earth have great virtues but do not speak of it.
Having adopted the metaphor approach from traditional Chinese literature, the project uses the decorations to express the idealsand replaces complexity with simplicity, reflecting humanity in details. The exquisite colors and furnishings also convey the delicate beauty in this home environment.

旺庭公馆
WANGTING MANSION

项目名称　旺庭公馆
设　　计　华浔品味装饰
建筑面积　240 平方米
项目地址　扬州
主要材料　大理石、仿大理石砖、乳胶漆、墙纸
工程施工　华浔品味装饰

本案主要采用了现代简约的手法，色彩冷静、线条简洁明快，更注重空间共性的表达，运用具有一致性和规律性的表达手法来处理空间的体块关系、收口关系、光影关系。而后根据空间自身的背景和场所感，通过对比、协调、统一等设计手法进行二次创作，通过家具风格的定位、配饰的选择和装饰材料的质感这些显性的元素来塑造属于这一空间的独特气质。

The project mainly adopts modernsimple style techniques by using calmcolors, concise and bright lines and especially the expression of spatial commonality. The expression techniques of consistency and regularity is applied to deal with the spatial relationships, the closing relationship and the light and shadow relationship. Then, according to the background of the space, contrast, coordination, unification and other design techniques are added for re-creation through the positioning of furniture style, the selection of accessories and the texture of decorative materials, creating a unique temperament that belongs to this project only.

童话·孔雀和天鹅的游戏

FAIRY TALE, THE GAME OF PEACOCK AND SWAN

项目名称 童话·孔雀和天鹅的游戏
设 计 华浔品味装饰 / 林凯平
建筑面积 115 平方米
项目地址 福州
主要材料 蒙娜丽莎瓷砖、伊丽莎白大理石、有家整木定制、宜人地板、名轩窗帘
工程施工 华浔品味装饰

在大多数的人心中都有一个巴黎梦,卢浮宫的人文情怀,香榭丽舍的罗曼蒂克,交织成最美的梦境。透过风光流转的明媚微光,在隽永如诗如画的建筑中,在繁华似梦似锦的大街上,暖风微醺,塞纳河畔的独特风景,那是时光沉淀下来的法式浪漫。

The Parisian dream has enchanted many people.The humanistic mood of the Louvre and the romance of the Champs Elysees intertwined into agorgeous dream that travels through the gentle and beautiful scenery among the picturesque buildings in the prosperous and dreamlike street; the intoxicating warm breeze and the unique scenery on the banks of the Seine, it is the French romance that has settled down with the years.

在这个案例中，除了黄铜和大理石的相得益彰、交相辉映外，白色的几何线条、细脚元素、轻体量的家具等，都为整体颜值加分不少，让原本简洁的空间更添层次。在色彩运用上，设计师特意选择了优雅大气的米杏色作为主调。同时，湖蓝和高级灰的局部点缀，既不影响空间氛围，又可以起到画龙点睛的作用，空间气质瞬间提升。

Apart from the bronze and marble that brings the best out in each other, the white geometric lines, the thin furniture legs and lightweight furniture also increase the attractiveness of the original simple space. In the selection of colors, the designer deliberately chose the elegant beige as the theme tone. Meanwhile, some areas are embellished by lake blue and grey colors to add a finishing touch and enhance the temperament of the space without changing its atmosphere.

韵·律
RHYTHM

项目名称 韵·律
设　　计 华浔品味装饰
建筑面积 115 平方米
项目地址 广州
主要材料 蒙娜丽莎瓷砖、伊丽莎白大理石、有家整木定制、宜人地板、名轩窗帘
工程施工 华浔品味装饰

本案对空间进行重新分割，以达到业主的功能所需，色彩及材质突破了住宅中的惯用色。
简约中带有一丝丝复古感，复古中又带有一丝现代感，不求奢华，只想在忙碌工作一天回到家有一片属于自己的净土。

In this case, apart from re-dividing the space to meet the functional needs of the owners, the color and material break through the usual color in the house.
Simple with a sense of retro, retro with a sense of modernity, not luxury, just want to return home in a busy day to have a piece of their own Pure Land.

整个空间家居色彩在雅致灰调的基础上，撞入了干净明亮的一抹纯色，流畅利落的线条将厨房和餐厅连通。少不了的现代化电器设备，融入整体的同时也点缀了空间。

On the basis of elegant gray tone, the whole space home color collides with a clean and bright pure color, and fluent lines connect the kitchen and dining room. Necessary modern electrical equipment, integrated into the whole at the same time also embellished the space.

恒利·江悦明珠
HENGLI · JIANGYUE MANSION

项目名称 恒利·江悦明珠
设　　计 华浔品味装饰/泸州公司
建筑面积 148 平方米
项目地址 泸州
主要材料 石砖、实木柱体装饰、实木墙板、装饰壁纸、天然石材
工程施工 华浔品味装饰

不加过多装饰与约束的美式风格，散发出迷人的休闲式浪漫，既有着欧式风格的华丽与贵气，又剔除了更多羁绊，同时又能寻找文化根基新的怀旧，贵气而不失自在与随意。
浅色系的沙发、木质家具搭配一株绿色的植物，传递着简洁、舒适的气息，在淡黄色光线的映衬下，呈现出雅致浪漫的韵味。明亮的光线从宽大的窗户照进，让客厅空间显得宽敞大气。

The project is designed with American style without too many decorations and restraints, providing a charming casual romantic environment Functionality is the priority of simple. American style and is therefor perfect for those who pursues comfort and also gives consideration to elegance.
The light color couch and wood furniture are matched with a green plant, conveying a concise and cozy atmosphere as well as a tasteful and romantic flavour. Bright light shines through from the large windows, making the living room look spacious.

天马相城

TIANMA VILLAGE

项目名称　天马相城
设　　计　华浔品味装饰 / 李俊峰
建筑面积　150 平方米
项目地址　烟台
主要材料　仿大理石砖、爵士白大理石、实木饰面板
工程施工　华浔品味装饰

简约不等于简单，它是经过深思熟虑后经过创新得出的设计思路的延展，不是简单的"堆砌"和平淡的"摆放"，简约风格不仅注重居室的实用性，还体现出了工业化社会生活的精致与个性，符合现代人的生活品位。与传统风格相比，简约装修设计所呈现的是剔除一切繁琐的设计元素，用最直白的装饰语言体现空间和家具所营造的氛围，进而赋予空间以个性和宁静。

Simplicity does not equal to plain, unrefined design. It is an extension of original design and ideas originated from deliberate decisions. It is not a simple "stacking" and a plain "placing". The simple style not only pays attention to the practicality of the living space, but also expresses exquisiteness and individuality of industrialized social life which caters to the tastes of modern people. Compared with traditional styles, minimalism eliminates all the complicated elements, and uses the most straightforward decorative language to convey the atmosphere created by the space and furniture, endowing the space with its own personality and tranquility.

上府名园
THE HOUSE GARDEN

项目名称　皇庭丹郡
设　　计　华浔品味装饰 / 胡宜长
建筑面积　135 平方米
项目地址　无锡
主要材料　奥特曼大理石、新中源瓷砖、进口墙布、进口硬包
工程施工　华浔品味装饰

本案设计上遵循自然、不高大不突兀、不多不少的设计原则。没有浮夸的地面拼花、没有精雕细琢的家具，也没有凹凸有致的护墙板。设计手法上采用了灰色的地面，咖啡色的平板木饰面，高级灰的墙纸，勾勒出了整个块面的空间。在色彩上，大面以黑白灰为主色调，用橙色和金色点燃了整个空间的激情。设计作品描绘了现代都市人的轻奢，某种程度上来讲，轻奢是多元化思考下的产物，符合了现代人的生活方式和生活态度，体现出了高效与时尚并存。

The design is totally natural, not lofty nor abrupt, not too much nor too little. There were no ostentatious floor parquet, no over-decorated furniture, no curved siding. The simple gray ground, the coffee-colored flat wood veneer, and the high-grade gray wallpaper outlined the space structure. Black, white and gray are applied as the basis, while orange and gold light up the passion of the whole space. This design work represents the easy attitude of modern urbanites with luxury. To some extent, casual Luxury as a style, is the product of diversified mode of thinking, and the synthesis of efficiency & style, which conforms to the lifestyle of modern people and their attitude towards life.

异 境
WONDERLAND

项目名称	异境
设　　计	华浔品味装饰 / 马千里
建筑面积	140 平方米
项目地址	长沙
主要材料	百叶木栅格、艺术涂料、瓷砖、木地板
工程施工	华浔品味装饰

家有千万种形态，可温馨，可沉稳，可时尚，可张扬……

本案业主是一位非常年轻的创业青年，找到我们的时候，便希望我们能给到他一个特别一点的家。本案天花设计大面积使用异形木百叶，地面铺贴也采用瓷砖和木地板异形拼接，地面与天花相呼应，解构了天圆地方的思维，让空间多了一种表现方式。虽是各种异形元素衔接，看上去却显得自然流畅，突破了形同样板一般不具表情的空间设计。这是将年轻创业者敢于冒险、勇于尝试新事物、富有创造力的品质融入到空间设计中的大胆实践，营造出契合业主自身特质的设计。

There are thousands of home styles available, such as warm, calm, stylish, high-profile and etc. The owner of this project is a very young entrepreneur, hoping that we can give him a special home. Special shaped wooden blinds cover a large area of the ceiling. Also special shaped tiles and wooden floors are pieced together on the ground, to echo the ceiling, symbolizing the ancient pattern of orbicular sky and rectangular earth. Although it is a combination of various exotic elements, it looks natural and smooth, breaking away from the ordinaryformula shapesin space design. It is a bold practice incorporating the adventurous, courageous and creative characters of young entrepreneurs intodesigning, creating an environment that fits the owner.

本案作品取名叫"异境",不仅是为了与空间中大量使用的异形设计元素相衬,也是为了展现屋主的前卫和不落窠臼的审美情趣。

开放式大阳台、去除隔断式厨房、采用玻璃隔断的形式,让书房呈现出半敞开半封闭状态,打开了空间布局,使得视觉观感达到最大化。

在材质选择上,木百叶、胡桃木及水泥灰的地面和墙面勾勒出质朴的美感,不同材质的衔接及比例分割使得空间协调自然。木材的温润与水泥的清冷,彼此平衡。

The project is named "wonderland", not only because of the abundant elements used, but also to show the avant-garde and unconventional aesthetic taste of the owner.

Open balcony and kitchen, andthe glass partition that help create the semi-open study, open up the layout tomaximize the visual space.

In the selection of materials, thewood louver, walnut and cement ash floor and wall finishsketch a simple beauty in the space.And the connection and proportion of different materials are harmonious. The warmness of the wood and the coolness of the cement balance each other.

FUSION CITY RESIDENTIAL AREA

融景城

项目名称 融景城
设　　计 华浔品味装饰 / 朱熙茜
建筑面积 130 平方米
项目地址 重庆
主要材料 硅藻泥、仿古砖
工程施工 华浔品味装饰

业主是自由职业者和公务员，他们喜欢温馨休闲又带点个性化的简美风格。简美风格整体给人一种简约大气的气息，没有过多复杂的空间布局，每处都格外干净利落，而且现代实用，无论是房顶的吊灯，还是脚下的地板，都没有琐碎多余的设计，表面光滑柔和，让人感觉温暖而舒适。

The owner prefers a cozy simple American style with its own characteristics. Without complicated spatial layout, each space is neat in appearance and practical in function. No redundant design is seen in this project, from the pendant lamp to the floor, but the smooth and soft surface texture creates a warm and comfortable living environment.

CREATIVE SPACE
创意设计空间

2020 / 品味TOP100

272 /	陈公馆 CHEN'S MANSION		314 /	宝能·太古城 BAONENG CELEBRITY GATEWAY
276 /	静 TRANQUILITY		316 /	星汇御府 SPARKLING STARS
278 /	著美·印记 BEAUTY·IMPRINT		320 /	东情西韵·融合风 EASTERN WARMTH WITH WESTERN CHARM · FUSION
282 /	晨 曦 MORNING SUNLIGHT		324 /	凤求凰 PHOENIX SEEKING HIS MATE
284 /	云水间 BETWEEN CLOUDS AND WATER		328 /	瑞和公馆 RUIHE HOMES
288 /	粹·白 PURE WHITE		330 /	西峰玖墅 JIUSHU VILLA
290 /	简·壹 SIMPLICITY · ONE		334 /	盛和花园 SHENGHE GARDEN
294 /	英伦豪城 BRITISH GARDEN		336 /	丽雅大院 GRACE COURTYARD
296 /	宁静于心 INNER PEACE		340 /	现代主义 MODERNISM
300 /	金德丰·会所 JINDEFENG · CLUBHOUSE		342 /	万科璞悦湾 VANKE PUYUE BAY
304 /	度·和 TOLERANCE HARMONY		346 /	云 舍 CLOUD COTTAGE
308 /	简·奢 SIMPLICITY · LUXURY		350 /	寂静的海 THE SILENT SEA
310 /	日光斜 THE SUNLIGHTS			

陈公馆
CHEN'S MANSION

项目名称 陈公馆
设　　计 华浔品味装饰
建筑面积 1000 平方米
项目地址 广州
主要材料 百叶木栅格、大理石、锐丽灯饰、缔美软装
工程施工 华浔品味装饰

本案借鉴传统建筑、造园手法提炼、整合空间，向传统生活方式的唯美致敬，在当代生活中重新寻找传统大宅府邸居住空间的气势之美、礼仪之美、舒适之美、雅致之美，从中延续传统意义的价值与观念。

在本案的设计过程中，设计师比较注重建筑风格的层次，理解空间与空间的关系，追求整体空间划分的独立性及合理性，做出空间层次感；梳理主人在空间的生活流程、秩序；增强情感交流，体现出节奏感。

This project draws lessons from traditional architecture and gardening techniques to refine and integrate the space, and salutes to the aesthetics of the traditional lifestyle to regain the beauty of traditional mansions, etiquette, comfort and elegance, and carry on the traditional values and concepts.

The designer pays attention to the layering in architectural style, by understanding the relationship between different spaces and pursuing the independence and rationality of the overall layout, creating a spatial hierarchy that helps to maintain an orderly daily life process and enhance emotional communication.

没有雕梁画栋，没有圆梁木雕，没有震眼大红，有的是对称结构和书香风韵，还有禅意飘香，设计秉承传统又体现现代，勾勒出空间典雅气质，营造出具有传统文化氛围的生活空间。

There is not any carved beam, painted rafter, wood carving or bright red in this space, instead there is symmetrical structure, scholarly charm and Zen flavor. The design inherits the tradition and reflects the contemporary, portrayingan elegant living space with a cultural atmosphere.

静
TRANQUILITY

项目名称 静
设　　计 华浔品味装饰 / 刘文林、罗建建、朱志健
建筑面积 110 平方米
项目地址 龙岗
主要材料 护墙板、黑镜、大理石、抛光砖、不锈钢、硬包
工程施工 华浔品味装饰

整个平由原始的两房，改成了现在的三房，结合客户的需求，给予客户更舒适，更适合的居住环境。客厅空间一点也不拥挤，浅色的沙发和深色的座椅搭配，沉稳又简单。现代台灯，刺绣抱枕，以及沙发后方墙上的泼墨画，让整个空间显得古色古香又与时俱进。沙发背景墙的设计加入了不锈钢、硬包，让整个设计更具时尚感。电视背景墙在设计手法上运用了取长补短的设计方式，让整个本身不算很长的背景墙，显现得更加醒目、对称。

We changed the layout from two bedrooms to three bedrooms to cater to their needs and provide them with a more comfortable living environment.

The living room is not spacious but does not appear crowded at all. The light color couch is matched with dark color chairs, looking calm and simple. The modern table lamp, embroidered pillows, and splashed-ink painting on the wall behind the couch give the space an antique and contemporary feeling at the same time. The wall behind the couch is covered with stainless steel finishing that adds a touch of fashion in the design. With skillful design techniques, the TV background wall appears more conspicuous and symmetrical.

项目名称	碧桂园凯茵梵登
设　　计	华浔品味装饰 / 毛志斌
建筑面积	810 平方米
项目地址	中山
主要材料	火烧原石、原木饰面、木地板
工程施工	华浔品味装饰

著美·印记
BEAUTY · IMPRINT

业主出生于文艺世家，强调传统文化中人与自然的和谐共处。所以在设计选材上也是尽可能地贴近自然的原木色材质。在入户绘画玄关和会客厅的贯穿设计中，墙面材质运用了比较粗犷的火烧原石和原木饰面材质，让整个空间更具有园艺的自然感。在各个不同的功能区域之间或隔断，或隔而不断，保证了空间的交流，又互不干扰。随处散落的植物，无不散发出一股自然的气息。

Born into a family of artists, the ownerattaches importance to the harmonious coexistence of man and nature in traditional culture and requires restoration of natural atmosphere in the living space.Therefore, natural wood color materials are selected.The walls of the entranceway and the living room arefinished withrough stones and wood veneer materials, filling the whole space with a natural feeling of garden. The different functional areas are partitioned, some areas are still connected visually tomaintain the communicationbetween areas without causing disturbance. The plantsscattered everywhere are all exuding a natural aroma.

房间以原木色为基调，最大的特色是采用原木色的饰面板装饰墙面，原汁原味的纹理隐约表达了现代人返璞归真的追求和对自然的向往。

素雅的色调，简洁的空间形式，这些赋予"高贵，典雅，舒适"一个全新的诠释。浅色木纹的修饰让空间有着和谐之美。

Based on the wood color, the most prominent feature in this space is the application of woodveneer to decorate the wall, the natural texture vaguely expresses the pursuit of returning to nature and the yearning for nature of modern people.

Elegant color, basic spatial form, graceful lighting, furniture andthe matching fabrics, all of which are giving a new interpretation of "noble, elegant and comfortable". The light wood color decorations bring a harmonious beauty to the space.

晨 曦
MORNING SUNLIGHT

项目名称	白鹭湖
设　　计	华浔品味装饰 / 林泽桐
建筑面积	150 平方米
项目地址	惠州
主要材料	黑色镜钢、木地板、大理石、护墙纸
工程施工	华浔品味装饰

主色调采用了贴近自然的原木色，淡淡的黄色跟清晨第一缕阳光一样，是如此的舒服、温和。原木的框架配上黑色镜钢的层板，软性温馨的木材与金属冷硬的碰撞，传统手法与现代中式的结合，奠定了整体稳重而又不失新颖的气息。男主人独爱的抽象意境墙绘，丰富了玄关、客厅、餐厅、主卧的格调。

The main color is a wood color that feels close to nature; the light yellow color isas cozy and gentle as the first sunlight in the morning. The wood frame is arranged with black mirror steel laminate, the soft and warm wood texture collides with the chill of the metal. The combination of traditional techniques and modern Chinese style has laid the foundation of a sedate and yet novel atmosphere. Additionally, the abstract artistic wall paintings picked by the owner embellish the entrance, living room, dining room and master bedroom.

云水间
BETWEEN CLOUDS AND WATER

项目名称 云水间
设　　计 华浔品味装饰 / 侯国富
建筑面积 1105 平方米
项目地址 佛山
主要材料 大理石、绿植、地砖、不锈钢、护墙板
工程施工 华浔品味装饰

绿地璀璨天城顶层宅邸，地产项目为超高层定位！揽入亚艺板块天际线，俯瞰四周，视野广阔。
设计方案意在将空间尽量打开，营造在云端，在水间的曼妙。
公共空间多以不同的材质、配饰、景色来区分，以达到一步一景的视觉效果。
现代高密度的城市建筑群里，能把自然引进室内，实属一种奢侈的享受。
无边际水景，与天际线融合，置身其中，立于天地之间，心胸旷达，万物了然于胸！

The plan is to open the space as much as possible to create a wonderland between clouds and water. Public space is usually distinguished by different materials, accessories and scenery in order to achieve a visual effect of "one step one view". It's a real luxury to bring nature into the room in modern high-density urban buildings.
Boundless waterscape integrated with skyline. Just as standing between heaven and earth, the mind is open and everything is so clear.

空间整体融合以当代的建筑风尚及东方美学的精神内核，旨在城市前沿的繁华静处，诠释一处契合城市雅贵内心之所向的现代东方人居空间，给予懂生活的人一处归心之所，寻得生活之本真含义。

The integration of contemporary architectural fashion and the spirit of oriental aesthetics creates a modern oriental living space for the city noble to find a tranquil place in the busy city and the true meaning of life.

PURE WHITE

项目名称 皇冠花园
设　　计 华浔品味装饰 / 林泽桐
建筑面积 200 平方米
项目地址 惠州
主要材料 板材、不锈钢、仿大理石砖、墙布
工程施工 华浔品味装饰

粹·白

本案以灰色为主色调，现代风格为基础，配以新中式软装做点缀，造型与色彩纯粹干净，简约又不失大方，正是业主"粹白"的育人准则。

全案采用了灰木纹的板材，搭配古铜色不锈钢边，大面积的平板设计，衬托出简洁干练的格调。女主独爱的抽象意境墙绘，映衬在沙发背景，过道背景与床头背景上，通过灯光的渲染，呈现出一番独特的意境。

With a dominant color of gray, the project is a modern style space embellished by neo-Chinese soft decoration. The shapes and colors are pure and clean, simple and sedate, which is exactly the owners' principle of the teaching principle "purified white".

The project employs gray wood grain board, bronze stainless steel edges and large panels tocreate a simple and elegant air. The abstract artistic wall paintingthat thewife adores, are arranged on the wall behindthe couch, the background wall at the aisle and bedside, presenting a unique artistic atmosphere through the lighting effect.

项目名称	三水别墅
设　　计	华浔品味装饰 / 罗韬
建筑面积	680 平方米
项目地址	乐从
主要材料	大理石、硅藻泥、大普家具
工程施工	华浔品味装饰

简 · 壹
SIMPLICITY · ONE

本案不流于传统元素堆砌的设计手法，保留传统中式风格的含蓄典雅，又融入新时代新元素，彰显现代品质感。用当代设计表达方式，将现代简约形式的中式家具及元素融入其中，使其成为一个具有"复古味道，现代美感"的传承中国文化的"灵魂"空间。

The design did not follow common techniques of using traditional elements, instead it preserves the subtle elegance of the traditional Chinese style, and incorporates new present-day elements to reflect the modern quality. Contemporary design techniques are applied to integrate the modern minimalist form of Chinese furniture and elements into the space, making it a "soul" space with a retro taste and modern aesthetic that inherits the Chinese culture.

空间色彩上，汲取墨色山水中的"灰度"为空间的主色调，稳重成熟；空间布局采取对称的形式，阴阳协调，加之具有古典元素的饰品摆件的点缀。整个空间内中式韵味浓烈，散发出亦古亦今的层次之美。

这方寸间的天地，待闲暇时刻，可手执一书，沏杯好茶，坐于茶台处，慢慢品味书香茗气，享受慢生活的惬意，享受当下的诗和远方。品茶为静养，健身为动养，在一片绿意盎然间，呼吸自然之灵气。

In terms of colors, the "grey" of the ink wash landscape painting is selected as the main color of the space, which provides a sedate and mature atmosphere. The spatial layout adopts a symmetrical form, to represent the coordination of the yin and yang. Embellished by the ornaments with classical elements, the whole space exudes a strong Chinese flavor and a combined beauty of the ancient and modern.

There is a world in this small space, when you are at leisure, you can hold a book, make a cup of good tea, sit at the tea table, and slowly taste the book and the tea, enjoying the slow-paced life, enjoying the poetry and the lofty. Tea is a nurturing in stillness, fitness is a nurturing in motion. You can breath the nature in this green land.

英伦豪城
BRITISH GARDEN

项目名称 英伦豪城
设　　计 华浔品味装饰
建筑面积 1242 平方米
项目地址 云浮
主要材料 微晶砖、诺贝尔大理石、中式家具
工程施工 华浔品味装饰

设计师在原建筑图上布置的主要功能进行了大胆的改造设计，原功能不够完善，不能满足目前业主的需求，设计师对原建筑功能进一步的深化设计。完善功能上做到"应有尽有"的原则，具备一定的商务功能（会客、休闲）。平面布局完整、大气，赋有贵族气质，杜绝虚假的奢华，体现真正的"高品质建筑"。功能与风格需紧密结合。设计时可按家庭成员的爱好、兴趣等设定为一个主题进行设计，反应真实、形象的生活场景。

The designer has been pondering over this question for a long time. He boldly redesigned the original building plan to meet the functional demands of the owners.
He improved the layout to ensure the space is fully functional including commercial purposes such as meeting and leisure activities. The layout is complete and sedate with a noble temperament, presenting a genuine "high-quality space". The styles of different area are also closely combined with the respective functions. The theme of each area is determined according to the family members' hobbies and interests to reflect the scenes of everyday life.

宁静于心

INNER PEACE

项目名称 万科水晶城
设　　计 华浔品味装饰 / 何志潮
建筑面积 810 平方米
项目地址 佛山
主要材料 大理石、实木地板、钨钢、艺术墙布、艺术涂料
工程施工 华浔品味装饰

空间就像一个舞台，每一个元素都相互配合，带给置身其中的人不一样的视觉体验和情感感受。本案运用利落的线条和整洁的块面，采用大理石、铝板、木栅格来营造通透感。同时采用设计师最喜欢的下沉式花园，来渲染一种大隐于市的淡泊与安然，每一个室内场所与核心庭院相对望，延续湿润雅致的生活气氛，日式枯山水的做法融合了抽象化的线条，阐释一方悠远意境。

The space is like a stage, the elements are cooperating with each other to bring various visual and emotional experiences to the residents living in it. This project is designed with neat lines and orderly block surfaces, using marble, aluminum panel and wood grid to create a semitransparent feeling. Meanwhile, my favorite type of underground garden is applied to portray a peaceful lifestyle indifferent to fame and fortune. The interior areas are facing to the central courtyard, continuing the warm and elegant atmosphere. And Japanese karesansui gardening techniques are integrated with abstract lines to interpret the profound artistic mood.

金德丰·会所
JINDEFENG · CLUBHOUSE

项目名称　金德丰·会所
设　　计　华浔品味装饰/刘秀扬
建筑面积　600平方米
项目地址　潮汕
主要材料　大理石、仿古砖、实木
工程施工　华浔品味装饰

自然、人与居所之间的和谐一直是设计探讨的终极问题，我们一直想在符合居住环境和适应业主生活方式之间找到平衡点，来设计理想的居所，在感受自然灵气的同时拥有一个素雅温馨的居所。因此我们将中式美学与现代元素融合，为这个幸福的家庭设计了饶有韵味的新中式住宅，让业主仿若置身户外，身心畅然。

The harmony between nature, people and residence has always been the ultimate questionin design. We have been trying to find a balance betweenthe living environment and the lifestyle of the owners to design the ideal residence and give the residents a cozy home as well as an enjoyment of the natural aura. Therefore we combined Chinese aesthetics with modern elements to build a neo-Chinesestyle residence for this happy family.

走进古色古香的中式特色家具装饰的房间，沉稳的历史厚重感，传统文化的怀旧情愫油然而生。儒雅文化的融入，东方韵味的传承立即呈现。本案例在整体风格上设计师把握了大气和稳重的家居氛围，采用太师椅、屏风、窗花、古画等中式元素，彰显出一种内敛而高贵的风范。

Upon entering this space decorated with antique Chinese-style furniture, the strong historic feeling and the nostalgia of traditional culture come to light. The integration of culture and oriental charm is presented immediately. The designer laysagraceful and sedate atmosphere in the overall space and employs Chinese elements such as wooden armchair, screen, window grille and ancient paintings to portray a restrained noble air.

TOLERANCE HARMONY

项目名称　度·和
设　　计　华浔品味装饰 / 邓会东、张雪
建筑面积　700 平方米
项目地址　惠州
主要材料　米洛西瓷砖、锐驰家具、硬包、不锈钢
工程施工　华浔品味装饰

度
·
和

君子量不极，胸吞自川流。
不争，和为贵。
本案在空间与功能体现上，主要以开放式，多动线去策划布置，完成本案的一个"和"。
在立面与用材上表达主题的"度"，立面造型上，没有太多的框，四方延升，不必拘束。
石材与木材的撞击，一硬一软。黑色与白色，一深一浅。或许凡事都有碰撞，有度量则万事和。

There is no limit to the amount of gentleman's tolerance which can harbor a hundred rivers.Do not scramble, it is harmony that is prized.

The project is mainly designed as an open space with multi traffic patterns to fulfill the functions, completingthe theme "harmony".

The theme of "tolerance" is expressed via the vertical façades and materials, there are not too many frames in the wall finishing, so the four sides can be extended without restraints.

Thestone and wood collides in the space, hard and soft, black and white, deep and shallow. Perhaps everything has a collision, and a gentleman's tolerance helps topreserve the harmony.

简·奢
SIMPLICITY · LUXURY

项目名称　太湖锦绣园
设　　计　华浔品味装饰 / 黄江林
建筑面积　700 平方米
项目地址　无锡
主要材料　大理石、护墙板、喷画、水晶
工程施工　华浔品味装饰

优雅的白色，通过灰色和浅咖啡色为颜色主线打造优雅简欧气息空间；空间的护墙及线条让空间的质感细腻、极富心思的家具配饰，隐约地显露了优越的品味。整个空间能让人感觉娴静舒适，让高贵优雅之气弥散开来，呈现出别样的奢华。设计师通过完美的点线运用，精益求精的细节处理，带给业主和家人不尽的舒服感，从而为业主打造了情深韵远的轻奢主义生活，营造了一个豪华典雅与悠闲舒适并存的别墅之家。

A tasteful and simple European atmosphere is created through the graceful white, gray and light coffee colors. The wall finish and lines further refine the texture of the space, and the delicate furniture accessories subtly express the fine taste. The whole space exudes a calm and comfortable feeling, allowing the elegant atmosphere to spread, presenting a different kind of luxury. Through perfectpoint-and-line design and the exquisite details, the designer brings the family an exceptional comfort and a luxurious living space, as well as an elegant and leisure home.

THE SUNLIGHTS

项目名称　日光斜
设　　计　华浔品味装饰 / 侯国富
建筑面积　700 平方米
项目地址　佛山
主要材料　大理石、百叶木栅格、木地板、麻质软装
工程施工　华浔品味装饰

日光斜

竹里缲丝挑网车，青蝉独噪日光斜。

如何营造诗意与田园？是我们要考虑的！空间要闲适，要接近自然，少不了光的作用。为了光，我们大动干戈！原本的地下室是封闭的，为了能让地下室有自然光线，不惜把整个花园全部下沉，又重新架空起来！这样不仅让地下室充满了光线，又打造出一个全新的立体花园，营造出了上下双层空间，整体空间多出一个维度！

How to create a feeling of poetry pastoral style for the space? That's what we need to think about. To be leisure and close to nature, light is indispensable, and we fight for this. Since the original basement is a closed space, the whole garden has been sunk and be raised again as stilt floor in order to let the natural sunlight into the basement. This not only fills the basement with light, but also creates a brand new three-dimensional garden, so a double-layer space was made and one more dimension was added for the space.

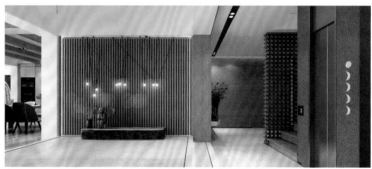

我们充分利用建筑的多面采光，把主卫生间的顶打开，让阳光洒进来，每天清晨，抬头便是夏蝉长鸣，日光倾泄！主卧室外自然的花园，故意把顶面打开一丝缝隙，让阳光和雨水洒下来，只有开门迎面吹来的和风，才能让人卸下一天的劳累，享受当下！

We made full use of the multi-faceted lighting of the building by opening the top of the main bathroom letting the sunlight in. Every morning, the summer cicadas is chirping, and the sunlight pouring out when looking up. The natural garden outside the master bedroom was intentionally opened a gap on the top to let the sunlight and rain fall. The gentle breeze will take off the tiredness of the day once opening the door. Just enjoy the moment!

宝能·太古城
BAONENG CELEBRITY GATEWAY

项目名称　宝能·太古城
设　　计　华浔品味装饰 / 洪文龙
建筑面积　280 平方米
项目地址　赣州
主要材料　木饰面、大理石、绒布、皮革、水晶
工程施工　华浔品味装饰

设计师力求打造栖居奢华而雅致的生活空间，高级灰的家饰与天然的大理石构建出一幅现代都市的生活情境。

高级灰被奉为经典的色彩，本案追寻沉稳雅致的高品质生活，却不过度繁复华丽，高级灰与大理石的白形成了强烈的对比，通过与几何灯具的搭配，渲染出奢华的情调。大理石墙面上的装饰画，一位舞者摆着舞姿，灯带缠绕着装饰画的四周，在空间中微微发光，与画面碰撞出极具韵律灵动的视觉冲击。

The project is designed as a luxurious and elegant living space. The graceful gray furniture and natural marble portray a modern urban life atmosphere.

Gray is regarded as a classical color. As this project pursues a calm and elegant high-quality life, the gray and the white of the marble form a strong contrast, rendering a luxurious ambience by matching with the geometric shape lamps. There is decorative painting on the marble wall, with a dancer dancing in it. A lighting belt decorated the frames of the painting, slightly glowing and generating a rhythmic and dynamic visual collision with the image on the painting.

星汇御府
SPARKLING STARS

项目名称　星汇御府
设　　计　华浔品味装饰 / 刘敏
建筑面积　280 平方米
项目地址　广州
主要材料　木饰面、大理石、绒布、皮革、水晶
工程施工　华浔品味装饰

色彩搭配："出挑的黄色不只是单一焦点，也在点缀天花、墙面上延伸和扩散着魅力。又为了挑起视觉美感，家具细节或装饰的框架更见功力。利用不同明暗、纯度、材质表达着黄色，使得主次分明，层次丰富。

Color matching: The yellow that stands out is not just a single focus, it is also dotted on the ceiling, extending on the wall finish and spreading its charm. The expertise can be seen in the details of the furniture or the decorative frames applied in order to provoke visual sense of beauty. Yellow is also an essential color in 2019 in the fashion world. The project uses different shades, saturation, and materials to express yellow, clarifying and marking the different layers.

玄关：将传统中式的提炼和升华，也是对传统中式的尊重，将东方美学用现代手法巧妙应用到空间中，对景方圆造型，地面祥云拼花，寓意吉言如意。把复杂的空间元素简练，给人一种沁润、意墨之境。将"四叶草"巧妙地植入对景造型，寓意幸运。

客厅：水而居，和山水融为一体，在含蓄节制的空间中感受自然的韵律，将大自然的美引入生活，寻找生活的本质，同时将禅意美学及极致装饰手法融汇其中，营造一种中国式的优雅奢华空间。

品茗区：自然暗香的木纹，在这里感受一茶一世界、一壶一人生东方优雅生活的高贵细腻。

Entranceway: Here it refines and sublimates the traditional Chinese style, which is also a respect for it. The oriental aesthetics is tactfully applied in the space with modern techniques. The auspicious pattern floor parquet, symbolizegood luck and happiness. The complicated elements are condensed to provide a refreshing and artistic atmosphere. Scenic focal point at the entrance, the "four-leaf clover" with an implication of good fortune is cleverly implanted into the scene, to lucky.

Living room: Integrated with the mountains and rivers, feeling the natural rhythm in the subtle and restrained space, introducing the beauty of nature into life, seeking the nature of life, and infusing Zen aesthetics and ultimate decorative techniques to create a Chinesestyle elegant and luxury environment. Beside the natural fragrant wood grain table, you can feel the oriental charm of a world in a tea and a life in a pot,

东情西韵·融合风
EASTERN WARMTH WITH WESTERN CHARM · FUSION

项目名称　棕榈泉独栋别墅
设　　计　华浔品味装饰 / 刘云
建筑面积　687 平方米
项目地址　重庆
主要材料　大理石、护墙板、墙布、仿真壁炉
工程施工　华浔品味装饰

空间规划不仅来自日常、"人本"的深刻观察，更有着去芜存菁后的内省与转化。不过分关注软装细节，更注重空间的逻辑与平衡关系，同时强调重质感、不繁复，线条组构利落而有层次，注重工艺细节与材质表现，比如大面积的护墙板、壁布的使用，材料的质感对于空间来说，就像高定服装的面料，会为家带来微妙的高级感。

Spatial planningcomes not only from daily, human-oriented observation, but also from introspection after removing the shackles. It pays more attention to the logic and balance of space and at the same timeemphasizes the texture and simplicity. The line structure is neat and layered, focusing on the details of the craft and materials, such as large-area wall panel, the use of wall covering. The texture of the materials for the space, like the fabric of haute couture, brings a subtle sense of high quality to the home.

一楼是公共区，客厅维持中空的原有结构，在几个维度都进行了对称化改造，营造大气端庄的空间气质。

二楼改造了走道的设计，打造了一个半弧平台，让两层楼间更具交互性。开放式厨房的设计，提供了足够的操作空间。三楼则是屋主的私人空间，房间和楼梯之间利用双开门做分割，房间内配置了卫生间、衣帽间加上和书房、阳台等功能空间，舒适性和私密性都可兼顾。

The first floor is a public area, the living room maintains the original hollow structure, and has been transformed symmetrically in several dimensions to create a dignified space.

On the second floor the aisle design is changed to create a semi-arc platform to increase the interactivity of the two floors. The open kitchen is designed to provide sufficient room for operation. The third floor is the private space of the owner. The bedroom and the stairs are partitioned by double doors. The bedroom is equipped with a bathroom, a closet, a study room and a balcony, taking both comfort and privacyinto consideration.

PHOENIX SEEKING HIS MATE

项目名称 凤求凰
设　　计 华浔品味装饰 / 王惠
建筑面积 170 平方米
项目地址 泉州
主要材料 水泥漆、仿大理石砖、乳胶漆、石材、黑色钛钢、木地板
工程施工 华浔品味装饰

凤求凰

今天这位家的主人，是位标准的 90 后，独立，拼搏，常年在外打拼，年纪不太却已然跻身于成功人士的圈层，后下文简称他为 T。

因为自身工作的原因，小 T 也时常在国外游历，对时尚，对生活，都有着自己独特的见解。我只和 T 做了最简单的沟通，在看了我两套作品后，T 在见面后发信息给我：按你的想法做。

The male phoenix returns to his home,

Flied over the four seas seeking his mate.

The owner is a typical Generation Y who was born in 1990s. Being independent and hard-working, he is already successful despite his young age. Because of his working and traveling experience at home and abroad, he has formed a unique understanding on fashion and life.

水泥漆，裸露的钢筋，黑色钛钢，暴力熊的大玩偶，夜空般黑色的餐厅顶面，许多不会在一般家庭出现的元素神奇地集结在一起。对于家具，最大的要求就是舒适度，懒人沙发，餐椅，床，甚至床品，都经过的精挑细选。我希望这个家不仅代表着主人的品味，更能陪伴着主人，承载着主人对于舒适生活和放松精神的需求。
希望这样的家，不仅 T 会喜欢，也能让他未来的媳妇满意。

Cement paint, bare steel bars, black titanium steel, big bear doll, and thenight-like black ceiling in the dining room many elements rarely seenin a home environment are magically arranged together here. For furniture, the principal requirement is comfort, therefore, the bean bag chair, dining chairs, bedsand even the bedding were carefully picked. I hope that this space will not only representthe taste of the owner, but also accompany the owner, carrying the owner's need for a comfortable life andrelaxation.

瑞和公馆
RUIHE HOMES

项目名称 瑞和公馆
设　　计 华浔品味装饰
建筑面积 140 平方米
项目地址 烟台
主要材料 仿大理石砖、护墙板、不锈钢
工程施工 华浔品味装饰

华灯初上，忙碌了一天的男主人回到家推开门映入眼帘的是一面整洁大气的大理石主题墙，在门厅把一天的疲惫归置好。客厅柔软的沙发在皮质背景的衬托下格外的温暖，期间夹杂的墙面顶面的金属修饰让眼球也有些许的兴奋。

孩子们在书房已经把功课温习完毕，妻子精心准备的晚餐已经就绪，餐桌上的美食在橙红色的背景的映衬下挑动着味蕾。洁白的大理石餐桌面，金色的桌腿，蓝色的柔软适中的餐椅，精心挑选的装饰品，无不让一天的疲惫消失殆尽。

This project is positioned as a high quality living space for an elite couple and their two children.

At the evening arrives, the busy husband returns home and upon opening the door, he is facing a clean and tidy marble wall. After leaving his exhaustion in in the entranceway, the soft couch in the living room appears exceptionally warm against the leather finish in the background. The slight metal finish at the top of the wall also catches his eyes.

The children have finished their homework in the study, and the wife has prepared a nice dinner which looks appealing against the orange-red background. And the white marble table, golden table legs, blue softdining chairs, and carefully selected decorations, have completely cleaned away the exhaustion.

西峰玖墅
JIUSHU VILLA

项目名称 西峰玖墅
设　　计 华浔品味装饰 / 胡戈
建筑面积 500 平方米
项目地址 扬州
主要材料 饰面板、玻璃、诺贝尔瓷砖、木地板
工程施工 华浔品味装饰

本案的业主是一个年轻的音响发烧友，自身对品质与时尚有敏锐的触觉和深刻的领悟。他喜欢前卫、简约又酷炫的事物，因此在室内设计中，主要围绕着大块面的黑白灰为线索而展开。在深入了解业主的诉求与喜好后，设计师发现二者志趣相投，从前期的构思到方案敲定，一气呵成，始终围绕着"简与酷"的调性，最终构成一个凝练纯粹的空间。

The owner of the project is a young audiophile who has a keen sense and profound understandingin quality and fashion. He likes avant-garde, simple and cool things, therefore the interior is mainly designed with large black, white and grey areas. After an in-depth communication on the owner's demands and preferences, the designer found that they shared similar hobbies. Eventually the design was completed in one go, from the concept to the finalization. It is all based on the "simple and cool" tonality, which ultimately constitutes a concentrated pure space.

盛和花园
SHENGHE GARDEN

项目名称　盛和花园
设　　计　华浔品味装饰 / 杨勇、徐丰文
建筑面积　190 平方米
项目地址　泰州
主要材料　硅藻泥、木地板、木质家具、护墙板
工程施工　华浔品味装饰

从一进门开始，素净的地面，整洁的沙发，纯纯的基调，很容易让人放松下下来，这样才能达到休息的目的。

家，顾名思义，就是一个充满温暖与温馨的地方，是每个人内心深处的眷恋，那一屡屡颜色一种种造型都会改变着我们的心情。简洁而不简单，成为大多数人们追求的方向。在郭先生的房子里，便有了充分的体现，从入户的清新，到卧室的温馨，完美地融为一体，形成一个轻松、愉悦的独立空间。

Hence from the entrance, the clean ground, the clean couch, the pure tone are designed to be relaxing, so that the owners can get enough rest.

Home, as the name suggests, is a place full of warmth and coziness. It is the strongest attachment deep inside, every color and style in it will affect our mood. Simple but not plain, has become the pursuit of most people, which is fully expressed at Mr. Guo's, from the freshness of the entrance, to the warmth of the bedroom, forming a relaxed and pleasant independent space.

丽雅大院
GRACE COURTYARD

项目名称	丽雅大院
设　　计	华浔品味装饰 / 赵淼、蔡天鳌、陈广
建筑面积	600 平方米
项目地址	宜宾
主要材料	大理石、浮雕、缔美软装、微晶石
工程施工	华浔品味装饰

业主是两位年轻人，喜欢比较浪漫一点的，觉得欧式过于复杂，现代过于简洁。于是，设计师强调浪漫的法式与现代结合，白色系为主色，祖母绿为点缀色。在硬装上面采用了简洁的法式线条，软装上面使用现代沙发。客厅除了法式美学应有的仪式感之外，运用现代极简具有线条感的现代家具，建立空间与人的情感链接。细节处的雕花、廊柱、线条，勾勒出古典美的精致和浪漫，卷草纹、罗马柱等法式经典元素则以更加精练的语言散发出妩媚优雅的情思。

The owners are two young people who prefers romantic style. Therefore our design team recommended a romantic combination of French and modern styles, with whiteas the dominant color and emerald green as the embellishment. Simple French style lines are used on the finishing, and modern sofas are applied in furnishing. In addition to the sense of ritual that French aesthetics contains, the minimalist modern furniture are chosen to create emotional links between the space and men The carvings, columns, lines portray the exquisiteness and romance in classical beauty, while the French classic elements such as scroll grass pattern and Roman columns exude an enchanting and graceful mood in a more concise language.

微风轻柔掠过，门厅左右两面的窗帘随风轻摆，吐纳着微醺的韵律，与一旁剔透的琉璃小品光影交汇，其中的情韵柔软而缱绻。

客厅在纯白的主基调上，用灰色威尼斯漆的岁月肌理与香槟金进行调和、碰撞，清丽的设色搭配着典雅的陈高艺术品、品质承袭的舒适家具，弥漫出一室轻盈雅致的古典风韵。

The breeze gently passes over, and the curtains on two sides of the hallway are swaying gently with the wind, breathing out a rhythm that is met withthe glaze next to them, revealing a soft and deep attachment.

Based on the color of pure white, the living roomblends the texture of gray venetian paint and champagne gold to create a contrast. The tranquil and beautiful hues are matched with tasteful artworkand comfortable fine furniture, filling the room with a light and elegant classical charm.

现代主义

MODERNISM

项目名称	雅居乐剑桥汇
设　　计	华浔品味装饰 / 吴微
建筑面积	300 平方米
项目地址	广州
主要材料	大理石、仿大理石砖、饰面板、不锈钢
工程施工	华浔品味装饰

设计师用黑、白、灰三色，打造了这一充满时尚与艺术感的现代风格空间。不累赘不繁复，利落的黑与纯净的白在空间游走撒欢，灰色则负责中和空间气质，带来安静的坚硬质感。灰白的背景墙减少了大面积黑色带来的压迫视觉感受。艺术性挂画成为视觉中心，使得整体风格既俏皮又现代。简洁有力的高级黑如同跳动的音符，将居室装扮得简练、硬朗、魄力十足。恢弘霸气间无限的设计灵感如期而至，为居室增添一抹纯粹与惊艳。简练硬朗的线条，霸气侧漏，仿佛装置艺术一般。

Designer uses black, white and gray to create this modern style case full of fashion and artistic sense without any burden and complexity.Clean black and pure white are wandering in space, and gray soften in between which creating a tranquil and hard feeling for the whole.Gray white background wall reduces the oppressive visual feeling brought by large area of black.Artistic hanging becomes the visual center, which makes the overall style both playful and modern. The simple and powerful high-level black is like a beating note, which makes the room simple, strong and full of energy. The design inspiration keeps coming just on time to add a touch of purity and grace to the room. The application of the black color makes both the living room and bathroom extraordinary , and the simple and strong lines are just as installation art.

万科璞悦湾

VANKE PUYUE BAY

项目名称 万科璞悦湾
设　　计 华浔品味装饰 / 唐远
建筑面积 150 平方米
项目地址 烟台
主要材料 定制橱柜、护墙板、玻璃、大理石
工程施工 华浔品味装饰

业主是一对母女俩，如何设计，做出什么样的效果已经在我脑海中呈现出来，回来以后慢慢实现，打造出客户想要的感觉。楼梯位置之前是放在餐厅位置，去现场以后第一感觉就是必须要调整楼梯的位置，空间一定要做开放式的，厨房、餐厅与客厅一定要开放，让家人之间得以拥有更多、更舒适的互动及活动空间。

The owners are a mother and a daughter, who are fashionable and havetheir own ideas. The project has an attic that can be used for storage. The stairs was originally placed in the dining room and has to be moved. The space must be open, the kitchen, the dining room and the living room must be To be open, to give more comfortable environment for interactions between the family members.

极简风格、干净利落的空间设计,摒弃了繁冗,留给我们更多思考的空间。低调的白灰色系沉稳大气,在快节奏的生活中让人静下心来,享受家中安静的美好。女孩房则采用当下流行的"脏脏粉"来衬托出业主小姑娘独有的少女气质。

Minimalist style, with its clean and neat design, abandoned the tedious decorations, leaving us more space for us to meditate. The understated white and gray is calm and sedate, enabling one to calm down in the fast-paced life and enjoy the quiet beauty at home. The daughter's bedroom adopts the current trendy dirtypowder to bring out the girlish temperament unique to her.

CLOUD COTTAGE

云舍

项目名称 云 舍
设　　计 华浔品味装饰 / 伍丽
建筑面积 470 平方米
项目地址 广州
主要材料 灰镜、宾格砖家瓷砖、原木家具、不锈钢
工程施工 华浔品味装饰

———

设计中我们放弃了常规的平面规划方式，改造的设计重点是打破各功能空间之间过分独立的关系，解决原有客厅、楼梯、厨房茶室的面积比例不足难以利用的痛点和难点。

在平面规划中为保证豪宅的视觉感受，我们打破了原建筑的格局，把客餐厅拼合成一个大空间，两层挑高的落地窗将阳光引入，空间流动着静谧和谐的气韵，取而代之的是 7 米 X7 米的大尺度开间，一进门就给人以宽敞舒适的感受；平衡各层空间功能比例的同时，调动区域气氛的活跃性，并充分利用轴线与艺术品摆件的关系，让动线更具体，渲染出整个空间的价值感与舒适度。

In the design of the duplex apartment, we gave up the conventional graphic planning. The design focus on breaking the excessively independent relationship between the functional spaces. The aim of the redesign is to solve pain point of insufficient space of living room, stairs and kitchen.

In order to ensure the visual experience of the mansion in the plan, we broke the original building pattern and combined the living room into a larger space. The 7 meters width and 7 meter height floor-to-ceiling windows allows the sunlight into room directly with quiet and harmonious atmosphere. While balancing the functional proportions of each areas, the design fully mobilizes atmosphere. Making full use of the relationship between the axis and the artwork, the value and comfort of the whole space is rendering clearly.

通过强化楼梯形态的序列性,对各层空间功能的串联与面积弥补,使各层空间有一定的互动性,让家人的互动性更强,使得家人之间更亲密。

By strengthening the sequence of the stair shape, the series and area of the spatial functions of each area could be compensated, so that each area has a certain interaction, which makes the family's interaction more powerful and makes the family more intimate.

寂静的海
THE SILENT SEA

项目名称 寂静的海
设　　计 华浔品味装饰
建筑面积 240 平方米
项目地址 珠海
主要材料 石材、仿古砖、玻璃、不锈钢、护墙板
工程施工 华浔品味装饰

"建筑以空间形式体现出时代的精神，这种体现是生动、多变而新颖的。"人们的居住空间，一直承载了人们的生活梦想，体现了不断变化的时代精神。

本案例选用多样的自然材质元素，以优雅的比例和材质色彩变化，坚持在材质选择和细节上的设计品位，在丰富多层次的灯光效果下，构建了一个具有高品质感和舒适感的生活空间。

"Architecture reflects the spirit of the times in a spatial form. This expression is vivid, changeable and novel." Our living spaces have always carried our dreams of life and embodied the changing spirit of times.

This project uses a variety of natural material elements.With elegant proportions and material color changes, it adheres to the taste in material selection and design details, building a life with high quality and comfort under the multi-level lighting effect.

大面积的深木色平静而柔和地散发出奢华的低调质感,与轻盈透气的白色石材呼应,整个空间有一种冷静的气质而又别具优雅。

在这样的空间行走,你仿佛心静到可以听到自己的呼吸声,面前如一片安静的海,放松而又令人愉悦。

The large area of dark wood color quietly and gently exudes an understated luxurious texture and echoes the light white stone to create a calm and graceful temperament. Walking in such a space, you seem to feel so calm thatyou can hear your own breathing, as if you are walking in front of a quiet sea, relaxed and pleasant.

图书在版编目（CIP）数据

品味TOP100 / 华浔品味装饰编著.—福州：福建科学技术出版社，2021.5

ISBN 978-7-5335-5483-5

Ⅰ.①品… Ⅱ.①华… Ⅲ.①室内装饰设计 Ⅳ.①TU238.2

中国版本图书馆CIP数据核字（2017）第288196号

书　　名	品味TOP100
编　　著	华浔品味装饰
出版发行	福建科学技术出版社
社　　址	福州市东水路76号（邮编350001）
网　　址	www.fjstp.com
经　　销	福建新华发行（集团）有限责任公司
印　　刷	中华商务联合印刷（广东）有限公司
开　　本	635毫米×965毫米　1/8
印　　张	44.5
插　　页	4
图　　文	356码
版　　次	2021年5月第1版
印　　次	2021年5月第1次印刷
书　　号	ISBN 978-7-5335-5483-5
定　　价	368.00元

书中如有印装质量问题，可直接向本社调换